THE AVIATOR'S SOURCE BOOK

THE AVIATOR'S SOURCE BOOK

BARBARA BUCHHOLTZ

St. Martin's Press | New York

For information, write: St. Martin's Press,
175 Fifth Avenue, New York, N.Y. 10010
Manufactured in the United States of America

Library of Congress Cataloging in Publication Data
Buchholz, Barbara.
Aviator's source book.
1. Private flying—Equipment and supplies—Catalogs.
2. Private flying—Miscellanea.
I. Title.
TL514.B8 629.133'029'4 81-21415
AACR2

ISBN 0-312-06252-4

CONTENTS

INTRODUCTION

The Aviator's Source Book was written and designed for general aviation enthusiasts, persons who are interested in all aircraft except military and commercial models. According to the most recent FAA statistics, it is a burgeoning group that now includes more than 827,071 pilots who fly more than 232,658 general aviation aircraft carrying more than 241,906,000 passengers each year. There are also many general aviation aficionados who will never take the controls of an airplane and who may never own their own airplane, but who still dream of doing so.

As with most hobbies and professions, there are innumerable items to purchase, some classified as essentials while others are merely extras that make a pursuit more pleasurable. The world of aviation is no exception. From the most serious, and expensive, purchase—the airplane—to the most whimsical—Snoopy models with airplanes for desk or bookshelf—there are hundreds of planes, accessories, services and educational materials. How much you fly, whether you fly for pleasure or for business, and how much you want to spend are all factors to weigh before buying.

The ideal way to shop is not by scurrying around to different manufacturers and distributors, but at home, by comparing the available equipment presented with photographs and specifications and performance sheets. But instead of your having to send off letters to hundreds of sources, you can have them all at your fingertips in this book.

We've done the work for you, compiling all the latest aeronautical equipment and accessories from more than 500 manufacturers, suppliers and miscellaneous sources. The information is presented in six chapters arranged by subject matter and subdivided into more specific categories by alphabetical order. We have listed addresses of manufacturers whom you can contact for additional information. Almost all prices are omitted, though, because we found that prices for aviation equipment, as with almost everything else, increase too rapidly, sometimes more than once a year. Most companies are more than willing, however, to provide you with up-to-date prices by mail or over the telephone.

Chapter One, "Learning to Fly," gets you off the ground and describes flight schools for all levels of expertise. Some offer weekend cram courses while others are year-round programs. You will also learn in this chapter about different flight simulators, and training aids such as books, computers and flight cases.

In the second chapter, "Aircraft," almost 200 different airplanes, from single-engine to multi-engine, and from propeller to jet, are described or illustrated. Easy-to-read charts list the manufacturer and model, number of seats, powerplant, fuel capacity, weight, cruise speed, rate of climb and other specifications and performance figures. Listed as well are companies specializing in modifying and converting aircraft. This chapter also provides helpful information on related subjects like title insurance, search, trading in an airplane or leasing one, and calculating operating costs.

Chapter Three details "Supplies for Maintaining

an Aircraft's Engine, Interior and Exterior," with those three topics subdivided for easy reference. Some supplies like rubber hoses and fittings are purely functional, while others are meant to offer high-in-the-sky comfort, such as sheepskin cushions and glassware with special aviation motifs and sayings.

Chapter Four, "Avionics," covers flight instruments, radio and radar equipment, and safety and emergency equipment.

Chapter Five, "Aeronautical Associations and Publications," will be of interest to both professionals and amateurs who like to browse through catalogs, books and magazines.

And finally, Chapter Six, "For the Flying Enthusiast," covers collectibles, gifts, clothing and jewelry which will please even the most sedentary aviator.

While we want to thank all of the sources listed in the book for their cooperation in providing us with detailed information and photographs, a special word of thanks is due to flying enthusiast Chet Kadish, who carefully and cheerfully reviewed the manuscript, offered suggestions for organizing the data and answered countless questions.

Needless to say, products change, and what was current at publication date may have become outdated by the time you read the book. We have done our best to provide the most comprehensive, up-to-date source book available.

B.B.B.
October 1981

1 / Learning to Fly

Pilots can hold a number of different FAA-approved licenses: private, commercial, flight instructor, ground instructor, and ATP. A pilot with a private license can fly and carry passengers for pleasure and convenience. If he or she, however, has a commercial license, the pilot can fly and carry passengers or freight for hire.

The type of aircraft the pilot can fly is also indicated on the license. For example, classifications include single-engine, multi-engine, land, sea, rotorcraft, and jet.

In addition, a pilot can add various ratings to the license, whether it is private or commercial. A pilot, for example, can add an instrument rating to a private license only after logging 200 hours and meeting the requirements of an instrument pilot.

Gaining a license and rating requires a pilot to first pass a written, oral, and flight examination administered by the FAA or designated examiners, or to graduate from a school that has been granted Examining Authority.

There are many flight schools across the United States that provide thorough training for the novice or the experienced pilot wishing to upgrade skills. The schools listed under "Flight Schools" range from weekend cram courses to four-year aeronautical universities.

Simulators also offer many advantages to the pilot learning to fly and to the experienced flyer trying to improve skills. The simulator monitors the pilot's performance and prints out the results.

There are three steps in simulation training: animated model boards containing indicators that react like those of an aircraft, a cockpit procedures trainer (CPT) simulating all aircraft systems, and finally the flight simulator itself, which provides the soundest training.

The capabilities of a simulator are extensive. The ATC-810 model, for example, provides general teaching skills such as basic instrument scan, orientation problems, wind correction angle, flight freeze model, takeoffs, and landings. Emergency and engine inoperative procedures include engine feathering and securing procedures, wing icing, fuel boost pumps inoperative, oil pressure problems, and fuel flow interruptions. Navigational capabilities include ADF tracking, VOR tracking, and many more.

While a full visual simulator sells for more than $2 million, there are desk-type instrument trainers for pilots who want to merely practice basic IFR procedures.

Additional training aids include books and manuals, instrument hoods, and records and cassettes.

FLIGHT SCHOOLS

AGS, INC.

AGS helps pilots prepare for their FAA written exams:

The Private Pilot written exam course, for both presolo and high time student pilots, includes instruction in navigation with plotter and computer, weather, regulations, and other subject areas on the FAA exam.

The Instrument Rating written exam course is designed for the private or commercial pilot about to start instrument flight training. The course includes explanations of IFR procedures and regulations, weather, approaches, and holding patterns.

The Airline Transport Pilot written exam course helps pilots reach the goal of becoming airline transport pilots.

AGS also offers Study Guides to prepare pilots for the weekend seminars: Private, Instrument, ATP, and Weather.

AGS weekend seminars are held in Atlanta, Georgia; Greenville, South Carolina; Miami, Florida; Oklahoma City, Oklahoma; Orlando, Florida; Plainview, New York; San Francisco, California; Tampa, Florida; Tulsa, Oklahoma; White Plains, New York.

Write for complete information:

AGS, Inc.
P.O. Box 43548
Atlanta, Georgia 30336

AIRLINE GROUND SCHOOLS, LTD.

The approximate price of Airline Ground School's three-day course is $185. Students can attend a class in a choice of cities around the country. Write for details on the three-day ATP Cram Course and three-day Flight Engineer Course.

Write to:

Airline Ground Schools, Ltd.
11050 Coloma Road, Suite U
Rancho Cordova, California 95670

AMERICAN FLYERS, INC.

Programs at American Flyers include the Professional Pilot Program (private, commercial, and instrument courses), which can be completed in four to five months; the Professional Instructor Program (certified flight instructor–airplane and certified flight instructor–instrument courses), which can be completed in just one to two months. The Multi-Engine courses usually require an additional two to four weeks.

The main campus for American Flyers is located in Ardmore, Oklahoma. Other campuses are located in West Chicago, Illinois; Fort Lauderdale, Florida; Santa Monica, California.

American Flyers, Inc.
Ardmore Airpark, P.O. Box 3241
Ardmore, Oklahoma 73401

AOPA AIR SAFETY FOUNDATION

The Instrument Flight Operations course gives an in-depth review of IFR flying practices in just two days. Subjects covered include instrument flight rules, regulations, procedures; ATC system, clearances, preferred routes, flight plans, filing; weather information system, coping with emergencies.

A sample of the daily schedule follows:

First Day
8:30 A.M. to 5:30 P.M.
—Introduction, registration, and objectives
—Review pilot and plane requirements
—Enroute charts, SIDs, and STARs
—Approach charts
—Advanced approach procedures
—Informal question and answer session

Second Day
8:30 A.M. to 4:00 P.M.
—Flight plans/planning
—Clearances
—Communications
—Holding and loss of communications
—Hazards of weather
—Professionalism

The Private/Commercial Pilot Written Examination course is for a student pilot who wishes to pass the FAA private pilot written examination. It is also recommended as a preparatory course for the FAA commercial pilot written examination, as a refresher course for current private pilots as well as former pilots who wish to become current, and as an introductory course to the Instrument Pilot/Instrument Flight Instructor written examination course.

The Instrument Rating/Instrument Flight Instructor Written Examination course is for the private pilot preparing for an instrument rating and for the experienced airman desiring to pass the Instrument Flight Instructor Rating written examination.

A sample of the daily schedule follows:

First Day
8:00 A.M. to 8:00 P.M.
—The IFR air traffic control system, procedures, and radio communications
—Use of IFR enroute and approach charts
—IFR navigation and flight operations on low and high altitude routes

Second Day

8:00 A.M. to 8:00 P.M.	—Aerodynamics —Aircraft instruments —IFR operations to and from high density airports —Aircraft performance and performance charts —Meteorology for IFR operations: weather theory, weather reports, forecasts and charts, elements of weather forecasting

Third Day

8:00 A.M. to 5:00 P.M.	—Federal aviation regulations —Airman's information manual —Aeromedical information —Review, practice examination, test-taking tips

The Airline Transport Pilot/Flight Engineer Basic Written Examination course prepares people fully for ATP certification.

The Flight/Ground Instructor Written Examination course trains a flight instructor for four FAA written examinations: Fundamentals of Instruction; Flight Instructor, Airplane; Ground Instructor, Basic; and Ground Instructor, Advanced.

In one day, students can learn about the atmosphere in motion, characteristics of air masses, fronts, icing, thunderstorms, wind shear, turbulence, restrictions to visibility, and much more. The course, Practical Aviation Weather, trains pilots of all experience levels to make intelligent decisions both before and during flight.

A sample of the schedule follows:

Hours

8:00 A.M. to 8:00 P.M.	—Principles of meteorology; general circulation, coriolis force, characteristics of high and low pressure areas, recognition of critical weather phenomena —Basic weather systems: air masses, fronts, jet stream —Weather modifying effects: terrain, seasonal, diurnal —Sources of weather information: reports, forecasts, charts —Application of weather information to the flight planning process —Preflight and inflight decisions —Actual sample flight problems and possible solutions: icing, turbulence, thunderstorm avoidance

The Updater course is for "rusty" VFR pilots who want to become refamiliarized with important basic aeronautical information and acquire the latest knowledge covering operation procedures, regulations, meteorology, and more.

AOPA also offers Flight Training Clinics for specialization in instrument procedures, mountain flying, survival training, and airmanship refresher training; Aviation Mechanic Refresher Courses to provide A1 and A&P mechanics with current information regarding maintenance, repair, and servicing requirements and recommendations originating from government and industry sources; and Flight Instructor Refresher Courses for certified flight instructors who wish to revalidate their certificates, attain their Gold Seal, or obtain professional refresher training.

Write for complete information and schedules to:

AOPA Air Safety Foundation
Air Rights Building
7315 Wisconsin Avenue
Washington, D.C. 20014

AVIATION TRAINING ENTERPRISES FLIGHT ACADEMY

The Professional Pilot Program is a 90-day program that includes private, commercial, and instrument ratings. The Professional Instructor Program, giving both CFI and A&I ratings, is a 30-day program. Requirements are that a person be at least 18 years old on the day of enrollment, be able to obtain a second-class medical and student pilot certificate, and be able to read, speak, and understand English.

Academy campuses are located in West Chicago, Illinois; Santa Monica, California; and Fort Lauderdale, Florida.

For more information, write to:

Aviation Training Enterprises Flight Academy
DuPage Airport
North Avenue (Route 64)
West Chicago, Illinois 60185
(312) 584-4700

Ft. Lauderdale Executive Airport
5500 N.W. 21st Terrace, Hangar 7
Ft. Lauderdale, Florida 33309
(305) 772-7500

Santa Monica Airport
3021 Airport Avenue
Santa Monica, California 90405
(213) 390-7878

AYRES CORPORATION

Ayres Corporation has a comprehensive school of agricultural aviation. The ground school consists of 126 hours of field and classroom training, covering subjects like field maintenance and repair, calibration, systems operation, and other subjects relating to aviation in general and ag-aviation in particular. In addition to aviation-related subjects, students receive instruction in entomology, chemical use, and safety and regulations related to aerial application.

There is also an agricultural pilot course that includes 50 hours of actual flying. Field entry and exit, swath runs, ag turns, flying under wires and around obstructions, as well as supervised solo spraying, are taught under actual agricultural conditions.

Ayres makes available a number of general-aviation courses: private, commercial, instrument, airplane multi-engine land (AMEL), flight instructor (ASEL) (CFI), flight instructor instrument (CFII), flight instructor multi-engine, and airline transport pilot (ATP).

For more information, write to:

Ayres Corporation
P.O. Box 3090
Albany, Georgia 31706

BRANTLY-HYNES

Brantly-Hynes offers helicopter training as its primary function. For recent graduates of the company's commercial helicopter pilot program, there is a rotorcraft helicopter flight instruction program. Helicopter instruction is given in Brantly-Hynes models B-2 and 305.

The factory flight school also works with pilots who wish to obtain advanced airplane pilot ratings. Cessna and Piper aircraft plus IFR simulators are used for the airplane courses.

Write for complete information to:

Brantly-Hynes
Box 697
Frederick, Oklahoma 73542

EMBRY-RIDDLE AERONAUTICAL UNIVERSITY

Embry-Riddle has a flexible academic calendar: a trimester system in which there are three equal terms of 15 weeks each. By attending year-round, a student can complete the four-year Bachelor of Science degree program in 32 months. There are seven general areas of aviation in which a student may receive a degree. All of the flight and maintenance programs of the university are approved by the Federal Aviation Administration.

For complete information, write to:

Embry-Riddle Aeronautical University
Star Route, Box 540
Bunnell, Florida 32010

EMERY SCHOOL OF AVIATION

Emery School of Aviation offers two professional pilot programs: the fixed-wing professional pilot program and the helicopter professional pilot program. A special part of the instructional experience includes mountain flying in the Rocky Mountains.

Entrance requirements include: being able to read and speak English; passing a second-class physical examination given by an FAA-designated medical examiner; and being a high school graduate or having an equivalency degree.

For more information, write to:

Emery School of Aviation
661 Buss Avenue
Greeley, Colorado 80631

FEDERAL CARRIERS

Federal Carriers offers flight instruction, aircraft maintenance, and charter service.

Prerequisites for enrollment in the Instrument Rating Course are a current FAA private/commercial pilot certificate; being able to read and speak English; an appropriate and current FAA medical examination; a high school diploma or equivalency degree.

The ground school includes a minimum of 30 hours of instruction. The typical student will require a total of 100 hours (20 hours tutoring, 30 hours classroom, and 50 hours of supervised self-study) to complete the ground school. Subjects include advanced aircraft performance, advanced meteorology, night flying, oxygen, IFR enroute and approach charts, and briefing for oral examination.

Flight training includes a minimum of 35 hours dual instruction before graduation with emergency procedures, night flying, complex aircraft, VOR orientation, VOR tracking, VOR approaches, NDB approaches, ILS approaches, and cross-country navigation.

Prerequisites for enrollment in the Private Pilot Certification Course are a current FAA Class 3 medical examination; a high school diploma or equivalency degree; a valid student pilot certificate, and being able to read and speak English.

The course prepares students for completion of the FAA pilot flight and oral examination, for the FAA written examination, and for knowledge and proficiency requirements stipulated for graduation from Federal Carriers' course.

The ground school includes a minimum of 35 hours of instruction in such subjects like aerodynamics, flight maneuvers, flight computer, navigation, weight and balance, meteorology, night flying, medical factors, and general safety.

Flight training includes a minimum of 35 hours of instruction (20 hours dual and 15 hours solo) and 16 hours pre- and postflight briefing in preparation for the FAA flight test. Subjects are preflight, taxi and runup procedures, normal takeoffs and landings, short and soft field takeoffs and landings, traffic pattern procedures, ground reference maneuvers, stalls, cross-country flying, night flying, instrument flight, and emergency procedures.

For complete information, write to:

Federal Carriers
P.O. Box 29
White Lake, New York 12786

TONY FINE GROUND SCHOOLS

Tony Fine Ground Schools offers a complete two-day program to prepare students for the FAA written examination. Refresher courses are available for free to former students.

For information, write to:

Tony Fine Ground Schools
P.O. Box 832
Sky Acres Airport
Millbrook, New York 12545

GRUMMAN AMERICAN FLYING CENTER

Student pilots at Grumman average 48 hours of flight time to earn their licenses. Grumman students have a 95 percent success rate on the FAA written exam. The Grumman curriculum consists of five sections: basic maneuvers, solo maneuvers, advanced maneuvers (specialty takeoffs and landings, radio navigation, unimproved airport operations, advanced ground reference maneuvers), cross-country, and finally, a complete review of ground work and air maneuvers as preparation for the FAA private pilot flight test.

For more information, write to:

Grumman American Aviation Corporation
P.O. Box 2206
Savannah, Georgia 31402

HUGHES HELICOPTERS PILOT TRAINING CENTERS

The materials, commissioned by Hughes Helicopters for use by its rapidly growing nationwide system of helicopter Pilot Training Centers, consist of a student kit featuring a private/commercial helicopter manual, a 160-page workbook, a 152-page maneuvers manual, student record folder, private and commercial stage exams, final exams, answer sheets, FAA and FCC forms, and diploma. The helicopter used is the Hughes 300C. There are centers located throughout the United States, Canada, England, and Sweden.

For more information, write to:

Hughes Helicopters
Culver City, California 90230

NORTHEAST HELICOPTERS, INC.

Helicopter flight training courses cost approximately $40 per hour solo or $55 per hour dual and utilize the Robinson R22 helicopter, equipped with King Avionics. Courses include those for private pilots, commercial pilots, and flight instructors.

There are other courses available such as instrument and air transport pilot, which may be carried out in the Bell Jet Ranger (206).

For complete information, write to:

Northeast Helicopters, Inc.
Hangar Two, Ellington Airport
Ellington, Connecticut 06029

PACIFIC WING & ROTOR, INC.

Pacific Wing & Rotor is a complete aviation training facility. The program offers private through ATP training in both fixed and rotary wing aircraft.

For complete information, write to:

Pacific Wing & Rotor, Inc.
Executive Air Terminal
3605 East Spring Street, Suite 210
Long Beach, California 90806

ROSS SCHOOL OF AVIATION, INC.

For inflight instruction, Ross uses a large fleet of Cessna and Piper aircraft. Multi-engine training is given in Piper twins. Students at Ross can practice a complete range of flight conditions in the Ross flight simulators, which are equipped for either single- or twin-engine programs. Courses offered are Private Pilot, Commercial Pilot, Instrument Pilot, Multi-Engine Land, Flight Instructor Airplane, Instrument and Multi-Engine Flight, and Airline Transport Pilot.

Write for more information, including current costs and requirements.

Ross School of Aviation, Inc.
Riverside Airport
Tulsa, Oklahoma 74107

SCHWEIZER AIRCRAFT CORPORATION SOARING SCHOOL

Schweizer Soaring School offers package courses for the beginner who wants to learn to soar and for the power pilot who wants to take up soaring and add a private sailplane rating to his present license. Private soaring courses begin each Monday. Other courses begin on Monday and Wednesday.

For complete information, write to:

Schweizer Soaring School
County Airport, Box 147
Elmira, New York 14902

SIERRA ACADEMY OF AERONAUTICS

Sierra Academy is an academically oriented pilot and flight engineer training school that specializes in providing programs for full-time resident students. Courses offered include private pilot through airline transport pilot in single- and multi-engine airplanes and helicopters, plus type ratings and flight engineer programs in Boeing 707, 727, and 737 turbojet aircraft. The school also features light jet programs (Cessna Citation and Rockwell Sabreliner) and hopes to offer training for the B747.

The jobs most graduates attain fall into two categories: airline flight engineer or first officer, and flight instructor/charter pilot.

For complete information, write to:

Sierra Academy of Aeronautics
Oakland International Airport
Oakland, California 94614

SOWELL AVIATION COMPANY, INC.

Sowell trains in the Cessna 150, Cessna 172, Cessna 172 RG, Cessna Citation, Piper Cherokee 140 and Piper Cherokee Arrow 200. The school offers instruction for the beginning student pilot through the airline transport pilot, as well as Citation jet and seaplane ratings.

For complete information, write to:

Sowell Aviation Co., Inc.
P.O. Box 1490
Panama City, Florida 32401

FLIGHT SIMULATORS

ANALOG TRAINING COMPUTERS, INC.

Analog Training Computers manufactures several simulators:

ATC-610J is a desk-size simulator, appropriate for a private pilot to gain and maintain instrument proficiency in his home or office. Along with VOR, the 610J features the same DME, ADF, and ILS capabilities as any well-equipped aircraft.

ATC-610K is an enclosed simulator for the professional training flight operation. All instrument time is loggable. With the addition of the optional X-Y flight plotter 20 of the 40 hours required for an instrument endorsement can be flown on the simulator.

Flight Simulator. Analog Training Computer Inc.

ATC-710 Professional Flight Simulator

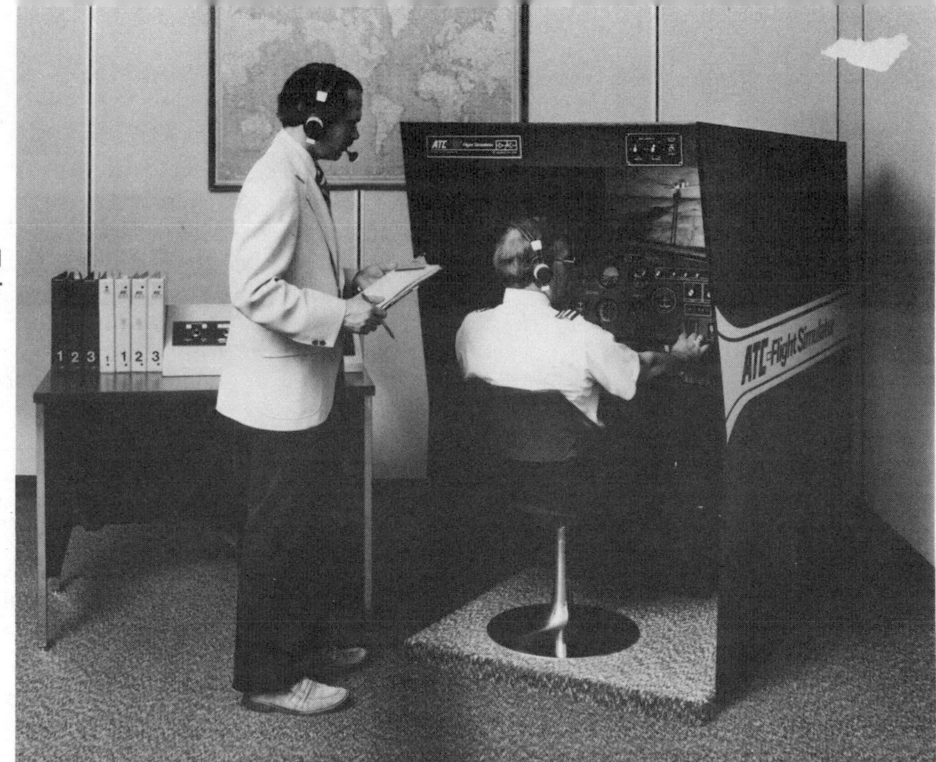

ATC-710 is sold in single- and multi-unit versions. Available features include RMI, digital DME, and new airspeed and vertical speed indicators.

ATC-810 is a twin-engine flight simulator and cockpit procedures trainer modeled after a turbocharged 6500-to-8000-pound cabin class twin. The ATC-810 includes navcoms, a transponder, ADF DME, VOR/ILS heads, audio panel, and an optional King 625 HSI. Digital LED frequency displays are standard.

ATC-112H is a helicopter IFR flight simulator, including the same IFR/VFR instrumentation as would be found in any well-equipped IFR helicopter.

Other products manufactured by Analog include:

ATC Monitor I enables a student to listen to flight assignment tapes on one player/recorder and record responses on the other. It teaches and helps a student improve his fluency in aviation communications.

ATC-112H Helicopter IFR Flight Simulator

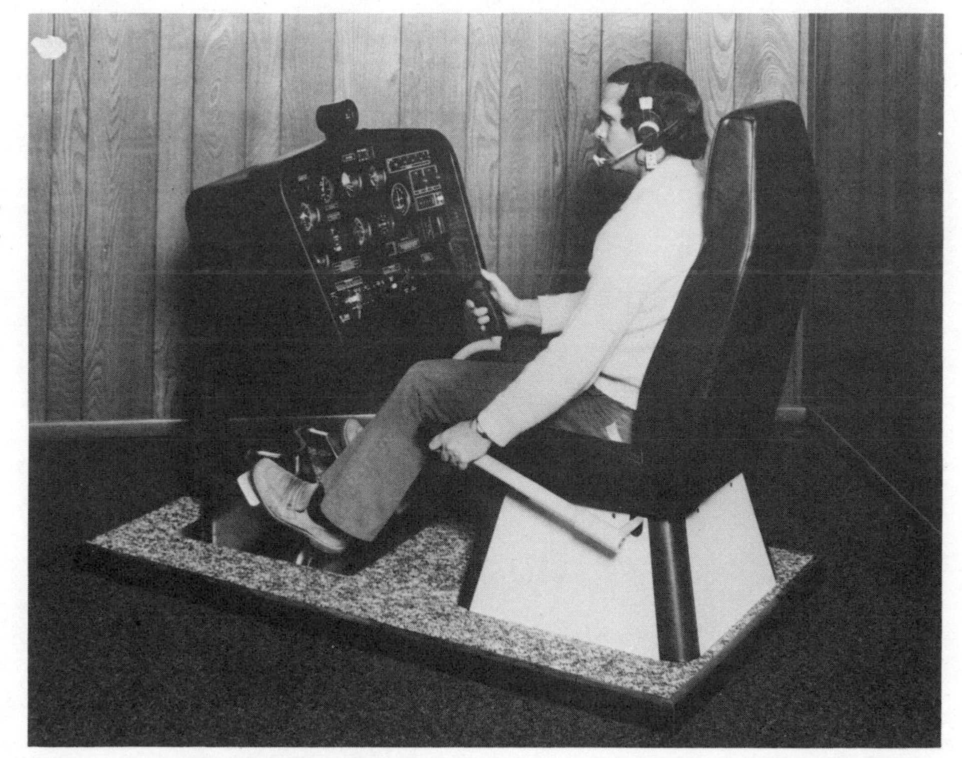

Flight Plotter provides a precise, permanent visual record of the simulator pilot's flight path. All flight influences such as headwinds, tailwinds, and drift are incorporated into the record.

For complete information, write to:

Analog Training Computers, Inc.
185 Monmouth Parkway
West Long Branch, New Jersey 07764

AVIATION SIMULATION TECHNOLOGY, INC.

Aviation Simulation Technology is a manufacturer of microprocessor-based single- and twin-engine flight simulators for major flight schools and fixed based operations. Preprogrammed memory cassettes are available covering areas approximately 200-by-200 nautical miles and containing all major navigation and airport approach facilities.

Models 300X and 201X are suitable for commuter airline, corporate, and flight school. These simulators consist of a flight compartment, an instructor's console, an external plotting aid, and a built-in digital computer with associated equipment.

Model 300 is the basic twin-engine simulator. Model 201 is the basic single-engine simulator. Both are designed for use with optional base and floor-mounted rudders and windshield enclosure.

For more information, write to:

Aviation Simulation Technology, Inc.
Hanscom Field-East
Bedford, Massachusetts 01730

FRASCA AVIATION

Frasca manufactures ground instrument trainers that can be adjusted to simulate the performance and response of an aircraft.

The Frasca 121 can have performance parameters to match almost any kind of single-engine aircraft. It includes a single switch for choosing fixed pitch or constant speed propeller and also includes retractable landing gear.

The Frasca 122 has an instrument panel that accurately reproduces the cockpit environment of a general aviation twin-engine airplane.

The Frasca 125H is an IFR helicopter flight simulator. It can perform all maneuvers encountered in normal instrument flight. Special additional capabilities include hover, translational flight, and auto rotation. Control appearance, movement, and responses are representative of the class and type of helicopter simulated. The collective lever, cyclic stick, and antitorque pedals perform the same functions as those in an aircraft.

The Frasca 300H/206 is another helicopter simulator that can perform all the maneuvers encountered in normal instrument flight. Special additional capabilities include hover, translational flight, and auto rotation. Control appearance, movement, and response are representative of the class and type of helicopter simulated. The collective lever, cyclic stick, and antitorque pedals perform the same functions as those in an aircraft.

The Frasca 300H/205 is a helicopter simulator. It can perform all maneuvers encountered in normal instrument flight. Special additional capabilities include hover, translational flight, and auto rota-

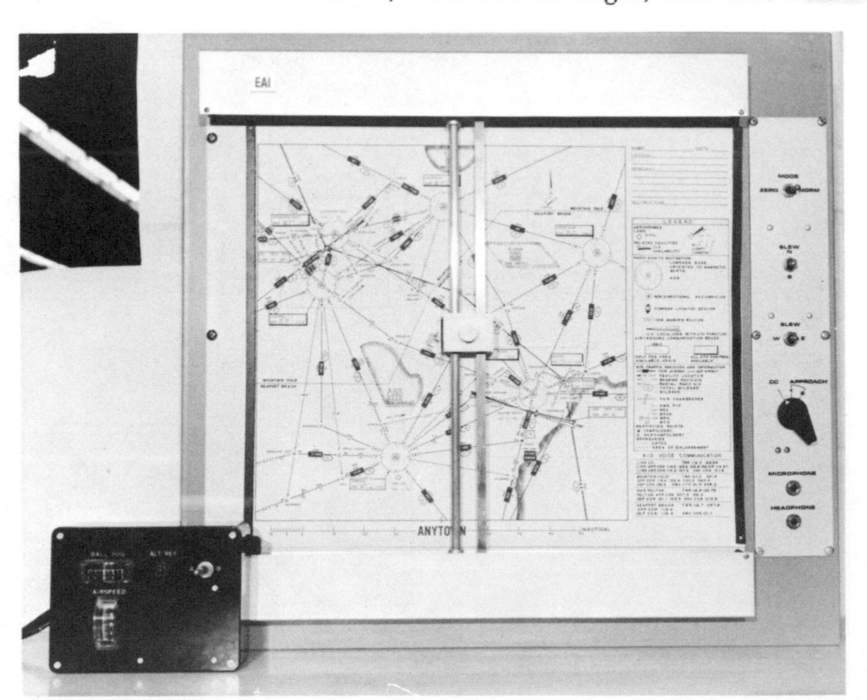

FRASCA 122 Flight Simulator

GAT-1 Recorder

tion. Control appearance, movement, and response are representative of the class and type of helicopter simulated.

For complete information, write to:

Frasca Aviation
606 S. Neil Street
Champaign, Illinois 61820

GENERAL AVIATION TRAINER

The standard GAT-1 trainer consists of a functional cockpit with controls and full flight performance, complete IFR capability, sound system, instructor controls, operations and maintenance manual, system diagrams, and detailed circuit description. The trainer includes an instrument flight package, an avionics system, and a malfunction insertion panel.

Simulated characteristics include pitch, roll, and yaw; engine torque and *P* factor; stall effects, oil temperature/pressure; fuel quantity; center of gravity; gross weight; outside air temperature.

GAT manufactures three trainers plus a helicopter operational trainer (HOT), which provides the basic flight/navigation and instrument training required for helicopters.

For complete information, write to:

General Aviation Trainer
Link Division, Singer Co.
Binghamton, New York 13902

PACER SYSTEMS

Pacer's wide range of trainers include single- and multi-engine systems as well as helicopter simulators.

Pacer Systems has assembled an audiovisual training package called the Pacer Mk II Courseware. The tapes are instructionally oriented and designed for playback with the listener seated at the Pacer Mk II controls. The documentation includes general

PACER MKIID Instrument Procedures Trainer

reference information and materials such as approach and terminal charts, pertinent to each lesson. An optional flight path recorder shows actual ground track of simulated flight path.

For complete information, write to:

Pacer Systems
87 Second Avenue
Northwest Industrial Park
Burlington, Massachusetts 01803

TRAINING AIDS

AVIATION MAINTENANCE PUBLISHERS

Aviation Maintenance Publishers puts out a series of technical books on aviation maintenance training as well as flight training aids. Following is a partial list of manuals:

Aviation Maintenance Handbook and Standard Hardware Digest—Order No. EA-AHS-1
Aircraft Inspection and Maintenance Records—Order No. EA-IAR
Applied Science for the Aviation Technician—Order No. EA-AS
Aircraft Hydraulic Systems—Order No. EA-AH-1
Aircraft Weight and Balance—Order No. EA-BAL
Aircraft Technical Dictionary—Order No. EA-ATD
Aircraft Reciprocating Engines—Order No. EA—ARE
Aircraft Gas Turbine Powerplants—Order No. EA-TEP
Aircraft Ignition & Electrical Power Systems—Order No. EA-IGS
Basic Electronics and Radio Installation—Order No. EA-BEM
Aviation Electronics—Order No. EA-AEG-1
Aircraft Instrument Systems—Order No. EA-AIS
Understanding Federal Air Regulations—Order No. EA-UFR
Federal Aviation Regulations for Aviation Mechanics—Order No. EA-FAR-1E
Aviation Maintenance Law—Order No. EA-AML-1
A&P Mechanics General Handbook—Order No. EA-AC65-9A
Preventive Maintenance for Pilots and Aircraft Owners—Order No. EA-AMP-1
Airframe Logbook—Order No. EA-AFL-1
Engine Logbook—Order No. EA-EFL-1
Radio Logbook—Order No. EA-ARL-2
Aviation Weather Services—Order No. EA-AC61-0045A
Private Pilot Flight Training Guide—Order No. EA-AC61-2A
Private Pilot–Combined Written Test Questions and Answer Book—Order No. EA-AC61-32C-1
Instrument Flying Handbook—Order No. EA-AC61-27B

For a complete list and other information, write to:

Aviation Maintenance Publishers, Inc.
P.O. Box 890
Basin, Wyoming 82410

AVIATION SUPPLIES & ACADEMICS, INC.

Aviation Supplies & Academics offers aviation training books. Some examples are:

Private Pilot Airplane Test
Instrument Pilot Airplane Test
ATP Airplane Test
Commercial Pilot Airplane Test
Flight Instructor Airplane Test
Cassette Instrument Ground School

For a complete list and other information, write to:

Aviation Supplies & Academics, Inc.
7201 Perimeter Road
Boeing Field
Seattle, Washington 98108

IOWA STATE UNIVERSITY PRESS

From Iowa State University Press come the following flight manuals:

W. K. Kershner, *The Student Pilot's Flight Manual*
W. K. Kershner, *The Advanced Pilot's Flight Manual*
W. K. Kershner, *The Instrument Flight Manual*
W. K. Kershner, *The Flight Instructor's Manual*
K. T. Boyd, *ATP: Airline Transport Pilot*
J. Webb, *Fly the Wing*

For current prices, write to:

Iowa State University Press
South State Avenue
Ames, Iowa 50010

AERO PRODUCTS RESEARCH

APR manufactures personal pilot equipment that makes a flight easier and safer. Items include computers, kneeboards, flight cases, checklist holders, pilot's flight desk, navigation plotters, sheets, and logbooks.

Some examples follow:

E6-B1—large plastic computer
E6-B9—large aluminum computer
CR-3—3¾-inch diameter, shirt pocket-size computer
CR-4—4¼-inch diameter computer
APR-204 T/S/D—traffic pattern orientator, aluminum computer
NPR-13—13-inch plotter
C-100—black (basic trainer) computer
0-100 VFR—all-aluminum kneeboard

APR also offers pilot training publications and FAA publications.

For a complete list and prices, write to:

Aero Products Research
11201 Hindry Avenue
Los Angeles, California 90045

JEPPESEN

Through dealers across the country, Jeppesen offers computers for solving navigation problems, plotters to simplify route planning, charts that show high density areas, vertical and lateral limits and airports, flight cases and manual binders, chart protectors, and navigation log/flight plan forms.

Write for a complete list:

Jeppesen
8025 E. 40th Avenue
Denver, Colorado 80207

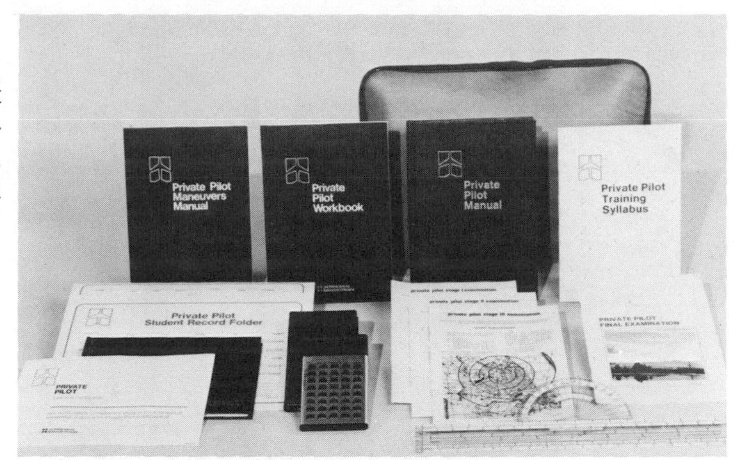

Jeppesen Sanderson Training Systems

MEASE ASSOCIATES

The flight manual *Pilot Talk, VFR,* teaches students and private pilots what phraseology to use and what frequency to call. Printed in large type, it allows just a two- or three-second glance at the text to let a pilot know how to talk like a professional would talk.

Write for current prices:

Mease Associates
74 St. Marks Avenue
Freeport, New York 11520

PILOT'S AUDIO UPDATE/EDUCATIONAL REVIEWS

A student or private pilot can listen and learn while driving or during any other convenient time and relisten as often as desired. A monthly cassette

Jeppesen's Airway Manual Service

Pilot's Audio Update

on tape brings up-to-the-minute flight information from government and nongovernment publications, refresher material, flight proficiency information, and other critical flying discussions.

For a current price list, write to:

Pilot's Audio Update/Educational Reviews
P.O. Box 582
Leeds, Alabama 35094

WJM SYSTEMS

Publications to improve a pilot's flying skills include *1–2–3 Simple Method to Smooth Landings; Crosswinds, Landings, Takeoffs, Taxiing;* and *You and Your Instructor.*

Write for additional titles:

WJM Systems
P.O. Box 201
Latham, New York 12110

2 / Aircraft

BUYING AN AIRCRAFT

Buying an airplane is an expensive undertaking, whether the plane is new or used. Title insurance provides protection against losses from previous title search errors, from failure to search for the title, from failure to properly record the plane with the FAA, from defective escrow procedures, and from breaks in the chain of the title.

AOPA Aircraft and Airmen Records Department provides a thorough search, disclosing the type of ownership and any recorded liens for which releases have not been received and recorded by the FAA. The Chain of Title shows the names and addresses of all recorded owners back to the date of manufacture. Normal service is to complete a title search within two to three working days after receipt. Rates for searches and insurance are available by request. Write to AOPA:

AOPA Aircraft and Airmen Records
Box 19244
Southwest Station
Oklahoma City, Oklahoma 73144

Insured Aircraft Title Service provides the proper procedures for selling or financing an aircraft, including a preliminary title search, appropriate escrow procedures, proper recording with the FAA, and issuance of a title insurance policy. A schedule of fees for aircraft title, escrow, and title insurance services is available upon request:

Insured Aircraft Title Service, Inc.
P.O. Box 19527
Oklahoma City, Oklahoma 73144

When buying, selling, or financing an aircraft abroad, contact International Aircraft Trusto Service at:

International Aircraft Trusto Service
One Maritime Plaza, 1313
San Francisco, California 94111

Once a title search has been satisfactorily completed, the following documents should be forwarded to the FAA Aircraft Registry, P.O. Box 25504, Oklahoma City, Oklahoma 73125: an Aircraft Bill of Sale Information and the Aircraft Registration Information, plus the Lender's Security Agreement or other evidence of lien.

If you are tired of your airplane and want another model, are looking to purchase a used airplane, or

☆ U.S. GOVERNMENT PRINTING OFFICE: 1978-773-353

FORM APPROVED
OMB NO. 04-R0076

UNITED STATES OF AMERICA

DEPARTMENT OF TRANSPORTATION — FEDERAL AVIATION ADMINISTRATION

AIRCRAFT BILL OF SALE INFORMATION

PRIVACY ACT OF 1974 (PL 93-579) requires that users of this form be informed of the authority which allows the solicitation of the information and whether disclosure of such information is mandatory or voluntary; the principal purpose for which the information is intended to be used; the routine uses which may be made of the information gathered; and the effects, if any, of not providing all or any part of the requested information.

The Federal Aviation Act of 1958 requires the registration of each United States civil aircraft as a prerequisite to its operation. The applicant for registration must submit proof of ownership that meets the requirements prescribed in Part 47 of the Federal Aviation Regulations.

This form identifies the aircraft being purchased, and provides space for purchaser and seller identification and signature. This is intended only to be a suggested bill of sale form which meets the recording requirements of the Federal Aviation Act, and the regulations issued thereunder. In addition to these requirements, the form of bill of sale should be drafted in accordance with the pertinent provisions of local statutes and other applicable federal statutes.

The following routine uses are made of the information gathered:

(1) To support investigative efforts of investigation and law enforcement agencies of Federal, state, and foreign governments.

(2) To serve as a repository of legal documents used by individuals and title search companies to determine the legal ownership of an aircraft.

(3) To provide aircraft owners and operators information about potential mechanical defects or unsafe conditions of their aircraft in the form of airworthiness directives.

(4) To provide supportative information in court cases concerning liability of individual in law suits.

(5) To serve as a data source for management information for production of summary descriptive statistics and analytical studies in support of agency functions for which the records are collected and maintained.

(6) To respond to general requests from the aviation community or the public for statistical information under the Freedom of Information Act or to locate specific individuals or specific aircraft for accident investigation, violation, or other safety related requirements.

(7) To provide data for the automated aircraft registration master file.

(8) To provide documents for microfiche backup record.

(9) To provide data for development of the aircraft registration statistical system.

(10) To prepare an aircraft register in magnetic tape and publication form required by ICAO agreement containing information on aircraft owners by name, address, N-Number, and type aircraft, used for internal FAA safety program purposes and also available to the public (individuals, aviation organizations, direct mail advertisers, state and local governments, etc.) upon payment of user charges reimbursing the Federal Government for its costs.

AC Form 8050-2 (8-76) (0052-00-629-0002)

FORM APPROVED:
OMB NO. 04-R0076

UNITED STATES OF AMERICA
DEPARTMENT OF TRANSPORTATION FEDERAL AVIATION ADMINISTRATION

AIRCRAFT BILL OF SALE

FOR AND IN CONSIDERATION OF $ THE UNDERSIGNED OWNER(S) OF THE FULL LEGAL AND BENEFICIAL TITLE OF THE AIRCRAFT DESCRIBED AS FOLLOWS:

UNITED STATES REGISTRATION NUMBER **N**

AIRCRAFT MANUFACTURER & MODEL

AIRCRAFT SERIAL No.

DOES THIS DAY OF 19
HEREBY SELL, GRANT, TRANSFER AND DELIVER ALL RIGHTS, TITLE, AND INTERESTS IN AND TO SUCH AIRCRAFT UNTO:

Do Not Write In This Block
FOR FAA USE ONLY

PURCHASER

NAME AND ADDRESS
(IF INDIVIDUAL(S), GIVE LAST NAME, FIRST NAME, AND MIDDLE INITIAL.)

DEALER CERTIFICATE NUMBER

AND TO EXECUTORS, ADMINISTRATORS, AND ASSIGNS TO HAVE AND TO HOLD SINGULARLY THE SAID AIRCRAFT FOREVER, AND WARRANTS THE TITLE THEREOF.

IN TESTIMONY WHEREOF HAVE SET HAND AND SEAL THIS DAY OF 19

SELLER

NAME (S) OF SELLER (TYPED OR PRINTED)	SIGNATURE (S) (IN INK) (IF EXECUTED FOR CO-OWNERSHIP, ALL MUST SIGN.)	TITLE (TYPED OR PRINTED)

ACKNOWLEDGMENT (NOT REQUIRED FOR PURPOSES OF FAA RECORDING: HOWEVER, MAY BE REQUIRED BY LOCAL LAW FOR VALIDITY OF THE INSTRUMENT.)

ORIGINAL: TO FAA

AC FORM 8050-2 (8-76) (0052-629-0002)

☆U. S. GOVERNMENT PRINTING OFFICE: 1979 - 674-825

FORM APPROVED OMB NO. 04-R0076

UNITED STATES OF AMERICA
DEPARTMENT OF TRANSPORTATION - FEDERAL AVIATION ADMINISTRATION
AIRCRAFT REGISTRATION INFORMATION

PRIVACY ACT OF 1974 (PL 93-579) requires that users of this form be informed of the authority which allows the solicitation of the information and whether disclosure of such information is mandatory or voluntary; the principal purpose for which the information is intended to be used; the routine uses which may be made of the information gathered; and the effects, if any, of not providing all or any part of the requested information.

The Federal Aviation Act of 1958 requires the registration of each United States civil aircraft as a prerequisite to its operation. An aircraft is eligible for registration only: (1) if it is owned by a citizen of the United States and it is not registered under the laws of any foreign country; or (2) if it is owned by a governmental unit. Operation of an aircraft that is not registered may subject the operator to a civil penalty.

This form identifies the aircraft to be registered, and provides the name and permanent address for mailing the registration certificate. The signature certifies United States citizenship as required by the Federal Aviation Act. Incomplete submission will prevent or delay issuance of your registration certificate.

The following routine uses are made of the information gathered:

(1) To support investigative efforts of investigation and law enforcement agencies of Federal, state, and foreign governments.

(2) To serve as a repository of legal documents used by individuals and title search companies to determine the legal ownership of an aircraft.

(3) To provide aircraft owners and operators information about potential mechanical defects or unsafe conditions of their aircraft in the form of airworthiness directives.

(4) To provide supportative information in court cases concerning liability of individual in law suits.

(5) To serve as a data source for management information for production of summary descriptive statistics and analytical studies in support of agency functions for which the records are collected and maintained.

(6) To respond to general requests from the aviation community or the public for statistical information under the Freedom of information Act or to locate specific individuals or specific aircraft for accident investigation, violation, or other safety related requirements.

(7) To provide data for the automated aircraft registration master file.

(8) To provide documents for microfiche backup record.

(9) To provide data for development of the aircraft registration statistical system.

(10) To prepare an aircraft register in magnetic tape and publication form required by ICAO agreement containing information on aircraft owners by name, address, N-Number, and type aircraft, used for internal FAA safety program purposes and also available to the public (individuals, aviation organizations, direct mail advertisers, state and local governments, etc.) upon payment of user charges reimbursing the Federal Government for its costs.

AC FORM 8050-1 (8-76) (0052-00-628-9004)

AC FORM 8050-1 (8-76) (0052-00-628-9004)

FORM APPROVED OMB NO. 04-R0076

UNITED STATES OF AMERICA
DEPARTMENT OF TRANSPORTATION - FEDERAL AVIATION ADMINISTRATION

AIRCRAFT REGISTRATION APPLICATION

CERT. ISSUE DATE

UNITED STATES REGISTRATION NUMBER N

AIRCRAFT MANUFACTURER & MODEL

AIRCRAFT SERIAL No.

FOR FAA USE ONLY

TYPE OF REGISTRATION (Check one box)

☐ 1. Individual ☐ 2. Partnership ☐ 3. Corporation ☐ 4. Co-Owner ☐ 5. Gov't.

NAME OF APPLICANT (Person(s) shown on evidence of ownership. If individual, give last name, first name, and middle initial.)

ADDRESS (Permanent mailing address for first applicant listed.)

Number and street: ____________________

Rural Route: P. O. Box:

CITY	STATE	ZIP CODE

☐ CHECK HERE IF YOU ARE ONLY REPORTING A CHANGE OF ADDRESS

ATTENTION! Read the following statement before signing this application.

A false or dishonest answer to any question in this application may be grounds for punishment by fine and/or imprisonment (U.S. Code, Title 18, Sec. 1001).

CERTIFICATION

I/WE CERTIFY that the above described aircraft (1) is owned by the undersigned applicant(s), who is/are citizen(s) of the United States as defined in Sec. 101(13) of the Federal Aviation Act of 1958; (2) is not registered under the laws of any foreign country; and (3) legal evidence of ownership is attached or has been filed with the Federal Aviation Administration.

NOTE: If executed for co-ownership all applicants must sign. Use reverse side if necessary.

EACH PART OF THIS APPLICATION MUST BE SIGNED IN INK.

SIGNATURE	TITLE	DATE
SIGNATURE	TITLE	DATE
SIGNATURE	TITLE	DATE

NOTE: Pending receipt of the Certificate of Aircraft Registration, the aircraft may be operated for a period not in excess of 90 days, during which time the PINK copy of this application must be carried in the aircraft.

AC FORM 8050-1 (8-76) (0052-00-628-9004)

merely want to sell your own plane, there are many good sources to turn to.

The most comprehensive is a classified newspaper, published out of Crossville, Tennessee, called Trade-A-Plane. In addition to a wide variety of planes for sale, the newspaper also lists aircraft tools for sale, hangars, radios, and other aviation-related information. Subscription rates and a subscription form can be obtained from:

Trade-A-Plane
Crossville, Tennessee 38555

Atlantic Aviation sells both aircraft and helicopters:

Atlantic Aviation Corp.
Greater Wilmington Airport
P.O. Box 15,000
Wilmington, Delaware 19850

Other good sources are the classified sections of major metropolitan newspapers and specialized business newspapers and magazines, such as the *Wall Street Journal*.

No matter how you decide to buy or sell an airplane, be sure to check such important facts as when the aircraft last had a major overhaul, what avionics are included in the purchase price, if any special features are part of the plane. Also be sure and take the airplane for a test flight.

Not everybody wants to own their own plane, but many individuals and businesses need to charter or lease an airplane, occasionally or on a regular basis. For these people, there are several services they can contact, including Atlantic Jet Charter, Hughes, Jet Fleet, Omni, and Xanadu.

Be sure to check carefully any lease plan to see that it meets your needs: number of years of the plan, deposit, monthly payments, purchase option, investment tax credit availability.

With headquarters in Wilmington, Delaware, and additional bases throughout the eastern and central United States, Atlantic Jet Charter provides efficient, comfortable business jet charters throughout the Western Hemisphere and Europe. The fleet includes the Westwind family, the Falcon 10, and other turbine-powered aircraft, all able to fly to 4,000 airports in the United States.

Atlantic Jet Charter's business aircraft are available 24 hours a day, 365 days a year. A number of charter options are available.

Atlantic Jet Charter
Greater Wilmington Airport
P.O. Box 15,000
Wilmington, Delaware 19850

The Hughes Executive Leasing Program (HELP) was started more than two years ago to tap the "industrial aid" market: corporations that could utilize one or more helicopters of the 500D class for transportation or fill a variety of other roles.

Hughes joined with the San Francisco leasing firm of Helicopter Capital Corporation to develop a leasing program and the result was the offering of three types of leases: leverage lease, operating lease, and finance lease. The actual payments on the three types of leases vary with the way the helicopter is outfitted.

Hughes Helicopter
Room 1005
1140 Connecticut Avenue NW
Washington, D.C. 20036

Jet Fleet Corporation operates a charter fleet of Fan Jet Falcons, Learjets, Sabreliners, and Cessna Citations from major cities throughout the United States. In addition to training and retraining its own crew members, Jet Fleet trains hundreds of pilots annually for many of the country's largest corporate aircraft operators.

The Jet Fleet flight control center operates 24 hours a day, making instantaneous trip price quotations, logging trip requests, and dispatching the appropriate aircraft. The center also provides the flight crews with weather information, takeoff and landing reservations, and other vital information. It will arrange ground transportation, hotel/motel reservations, conference rooms, special catering, and other customer requirements.

Other services offered by Jet Fleet include aircraft sales, maintenance and modification, warranty and repair station, avionics installations, interior modifications and installations, loaner pilot programs, and owner/assist programs.

Jet Fleet Corporation
P.O. Box 7445
Dallas, Texas 75209

The Omni International Jet Trading Floor, based in Washington, D.C., offers many services:

- Appraisals—Omni experts appraise aircraft
- Purchasing—Omni can and will buy any size jet or jet fleet on the spot
- Marketing—aggressive marketing makes possible efficient, economical aircraft sales
- Trading—aircraft trading is Omni's primary business for all types of business and commercial aircraft

- Leasing—Omni offers short, medium, and long leases and rentals
- Financing—Omni staff members include CPA financial experts and an inhouse tax counsel

Omni International Jet Trading Floor
P.O. Box 34449
Washington, D.C. 20034

Xanadu is an aircraft brokerage firm that provides new and used aircraft. The company presently has agents in 22 major cities united by a computer-linked network, much the same as real estate's Century 21.

Xanadu Aviation
Box 52-2224
Miami, Florida 33152

Once a person has purchased an airplane, he can gain substantial savings in taxes by keeping careful records. A book such as Taxlogs enables an airplane owner to record all of the flight and aircraft data required by the FARs on clearly designed log pages. The book also contains forms for documenting all possible aviation tax deductions, plus complete instructions for using the forms. The book measures 6-by-9 inches, has 128 pages, and is bound in a green hardcover.

Taxlogs Unlimited
20 Galli Drive
Ignacio, California 94947

In the following sections of this chapter, all speeds are indicated in miles per hour, all distances are measured in statute miles, and all gross weights are measured at sea level.

The worksheet below will enable an aircraft owner to calculate the total hourly operating cost of the plane, taking into account local operating conditions and prices.

DIRECT OPERATING COST

Consumables:	Fuel______GPH@______gal	$ __________
	Oil/hour	__________
Maintenance Labor:	Scheduled	__________
	Unscheduled	__________
Reserves for:	Airframe spares	__________
	Engine overhaul	__________
	Engine maintenance and spares	__________
	Major component overhaul	__________
	Retirement items	__________
	Total Direct Cost per Hour	$ __________
Fixed Operating Cost:	Depreciation	$ __________
	Liability insurance	$ __________
	Pilot salary (if any)	$ __________
	Hangar rental	$ __________
	Miscellaneous	$ __________
	Total Annual Fixed Costs	$ __________
Total Annual Fixed Costs Divided by Estimated Annual Flight Hours =		$ __________
	Direct Hourly Cost (from above)	__________
	Total Hourly Operating Cost	$ __________

Single-Engine Aircraft

Beechcraft	9709 East Central Wichita, Kansas 67201	Skipper Sundowner Sierra Bonanza V35B Bonanza F33A Bonanza A36 Bonanza A36TC
Cessna	P.O. Box 1521 Wichita, Kansas 67201	152 Hawk XP Skyhawk 180 Skywagon 185 Skywagon Skylane and Turbo Skylane Skylane RG and Turbo Skylane RG Stationair 6 and Turbo Stationair 6 Centurion, Turbo Centurion, and Pressurized Centurion Stationair 8 and Turbo Stationair 8 Cutlass RG
Lake Aircraft	Laconia Airport Laconia, New Hampshire 03246	Lake Buccaneer
Maule	Spence Air Base Moultrie, Georgia 31768	M-5 Lunar Rockets
Mooney Aircraft	P.O. Box 72 Kerrville, Texas 78028	201 231
Piper	Lock Haven, Pennsylvania 17745	Tomahawk Super Cub Warrior II Archer II Dakota Arrow IV and Turbo Arrow IV Saratoga and Turbo Saratoga Saratoga SP and Turbo Saratoga SP
Taylorcraft	15600 Commerce, N.E. P.O. Box 243 Alliance, Ohio 44601	F21
Varga Aircraft	12250 East Queen Creek Road Chandler, Arizona 85224	2150A 2180

Beechcraft Skipper

Skipper

The Skipper is a two-place, low-wing, all-metal airplane powered by a Lycoming 0-235-L2C engine rated at 115 HP. The Skipper includes left and right cabin entrance doors, foam seat cushions, tinted windshield and windows, windshield defroster, map stowage and cabin door locks and pulls.

Manufacturer and Model	Seats	Powerplant	Fuel Capacity (std/opt. gal)	Weights Gross/ Empty (lbs)	Cruise Speed (mph) 75% @ alt 65% @ alt	Optimum Range (m w 45 min. rsv) 75% @ alt 65% @ alt	Takeoff/ Landing Distance (ft. over 50′ obst)	Rate of Climb (fpm)	Service Ceiling	Stall Speed (gear/flaps down, mph)
BEECH	2	Lyc. 0-235-L2C	29	1675/	121 @ 6,500′	394 @ 6,500′	1280/	720	12,900′	54
Skipper		115 hp.		1100	112 @ 4,500′	425 @ 4,500′	1313			

Sundowner

A Lycoming O-360-A4K engine rated at 180 HP powers the Sundowner, a four-place, low-wing, all-metal airplane. The aircraft has a conventional three-control system with rudder pedals on the left side, individual toe-operated brakes, parking brake, and nose wheel steering.

Beechcraft Sundowner

Manufacturer and Model	Seats	Powerplant	Fuel Capacity (std/opt. gal)	Weights Gross/ Empty (lbs)	Cruise Speed (mph) 75% @ alt 65% @ alt	Optimum Range (m w 45 min. rsv) 75% @ alt 65% @ alt	Takeoff/ Landing Distance (ft. over 50′ obst)	Rate of Climb (fpm)	Service Ceiling	Stall Speed (gear/flaps down, mph)
BEECH Sundowner 180	4	Lyc. O-360-A4K 180 hp.	57	2450/ 1502	137 @ 8,500′ 124 @ 8,500′	613 @ 8,500′ 670 @ 8,500′	1955/ 1484	792	12,600′	58

The Sierra is a four-to-six place airplane with retractable gear, powered by a Lycoming 10-360-A1B6 fuel-injected engine rated at 200 HP and with constant speed propeller. Wing span of the plane is 32′9″, length is 25′9″, and height is 8′1″.

Sierra

Manufacturer and Model	Seats	Powerplant	Fuel Capacity (std/opt. gal)	Weights Gross/ Empty (lbs)	Cruise Speed (mph) 75% @ alt 65% @ alt	Optimum Range (m w 45 min. rsv) 75% @ alt 65% @ alt	Takeoff/ Landing Distance (ft. over 50′ obst)	Rate of Climb (fpm)	Service Ceiling	Stall Speed (gear/flaps down, mph)
BEECH	4-6	Lyc.IO-360	57	2750/	157 @ 10,000′	743 @ 10,000′	1660/	927	15,385′	69
Sierra 200		A1B6 200 hp.		1713	146 @ 10,000′	770 @ 10,000′	1462			

Beechcraft Sierra

Bonanza V35B

A four- or five-person, single-engine, high performance business and pleasure airplane with fully retractable tricycle landing gear, the Bonanza V35B is powered by a Continental six-cylinder 10-520-BB fuel-injected engine rated at 285 HP. The airplane is licensed in the utility category at maximum takeoff weight.

Manufacturer and Model	Seats	Powerplant	Fuel Capacity (std/opt. gal)	Weights Gross/ Empty (lbs)	Cruise Speed (mph) 75% @ alt 65% @ alt	Optimum Range (m w 45 min. rsv) 75% @ alt 65% @ alt	Takeoff/ Landing Distance (ft. over 50′ obst)	Rate of Climb (fpm)	Service Ceiling	Stall Speed (gear/flaps down, mph)
BEECH Bonanza V35B	4-5	Cont. IO-520-BB 285 hp.	74	3400/ 2117	198 @ 6,000′ 187 @ 8,000′	824 @ 6,000′ 894 @ 8,000′	1769/ 1324	1167	17,858′	58

Beechcraft Bonanza V35B

Beechcraft Bonanza F33A

Bonanza F33A

The maximum speed of the Bonanza F33A is 209 mph, with power provided by a six-cylinder 10-520-BB fuel-injected engine rated at 285 HP. Avionics that are standard equipment are a King KX-170B (720 channel) Com Transceiver, (200 channel) Nav Receiver with KI-208 VOR/LOV Converted-Indicator; microphone; headset; cabin speaker; NAV/GS and Com Antennas.

Manufacturer and Model	Seats	Powerplant	Fuel Capacity (std/opt. gal)	Weights Gross/ Empty (lbs)	Cruise Speed (mph) 75% @ alt 65% @ alt	Optimum Range (m w 45 min. rsv) 75% @ alt 65% @ alt	Takeoff/ Landing Distance (ft. over 50′ obst)	Rate of Climb (fpm)	Service Ceiling	Stall Speed (gear/flaps down, mph)
BEECH Bonanza F33A	4-5	Cont. IO-520-BB 285 hp.	74	3400/ 2132	198 @ 6,000′ 187 @ 8,000′	824 @ 6,000′ 894 @ 10,,000′	1769/ 1324	1167	17,858′	58

Bonanza A36

Powered by a Continental six-cylinder 10-520-BB fuel-injected engine rated at 285 HP, the Bonanza A36 has a maximum speed of 206 mph. Four hundred pounds of baggage can be stored in the aft cabin compartment; 70 pounds in the extended rear compartment. Special features include rear passenger double door, safe flight stall warning horn, quick removable gas tank caps, emergency locator transmitter, and complete exterior polyurethane paint job.

Beechcraft Bonanza A36

Manufacturer and Model	*Seats*	*Powerplant*	*Fuel Capacity (std/opt. gal)*	*Weights Gross/ Empty (lbs)*	*Cruise Speed (mph) 75% @ alt 65% @ alt*	*Optimum Range (m w 45 min. rsv) 75% @ alt 65% @ alt*	*Takeoff/ Landing Distance (ft. over 50′ obst)*	*Rate of Climb (fpm)*	*Service Ceiling*	*Stall Speed (gear/flaps down, mph)*
BEECH Bonanza A36	4-6	Cont. IO-520-BB 285 hp.	74	3600/ 2191	193 @ 6,000′ 181 @ 8,000′	802 @ 6,000′ 860 @ 10,000′	2040/ 1450	1030	16,600′	60

Beechcraft Bonanza A36TC

Bonanza A36TC

The engine of the Bonanza A36TC is a Continental six-cylinder TSIO-520-UB turbocharged, fuel-injected model, rated at 300 HP. The plane can seat four to six persons in a cabin measuring 12′7″ long and 3′6″ wide. More than 470 pounds of baggage can be stored in the aft and rear compartments.

Manufacturer and Model	*Seats*	*Powerplant*	*Fuel Capacity (std/opt. gal)*	*Weights Gross/ Empty (lbs)*	*Cruise Speed (mph) 75% @ alt 65% @ alt*	*Optimum Range (m w 45 min. rsv) 75% @ alt 65% @ alt*	*Takeoff/ Landing Distance (ft. over 50′ obst)*	*Rate of Climb (fpm)*	*Service Ceiling*	*Stall Speed (gear/flaps down, mph)*
BEECH Bonanza A36TC	4-6	Cont. TSIO-520 UB 300 hp.	74	3650/ 2269	223 @ 25,000′ 201 @ 18,000′	773 @ 25,000′ 800 @ 20,000′	2012/ 1449	1165	25,000′	65

CESSNA

152

The 152 Trainer, the most frequently ordered avionics and accessories package, is a standard feature of the 1981 model 152. Savings on a 152 Trainer, over the prices of the items if they were individually selected at list price, equals more than 20 percent.

Manufacturer and Model	Seats	Powerplant	Fuel Capacity (std/opt. gal)	Weights Gross/ Empty (lbs)	Cruise Speed (mph) 75% @ alt 65% @ alt	Optimum Range (m w 45 min. rsv) 75% @ alt 65% @ alt	Takeoff/ Landing Distance (ft. over 50′ obst)	Rate of Climb (fpm)	Service Ceiling	Stall Speed (gear/flaps down, mph)
CESSNA 152	2	Lyc. O-235-L2C 110 hp.	24.5/ 37.5	1675/ 1109	123 @ 8,000′ 115 @ 8,000′	368 @ 8,000′ 403 @ 8,000′	1340/ 1200	715	14,700′	50
CESSNA 152 Aerobat (A)	2	Lyc. 0-235-L2C 110 hp.	245/ 375	1675/ 1135	122 @ 8,000′ 112 @ 8,000′	360 @ 8,000′ 403 @ 8,000′	1340/ 1200	715	14,700′	50

Cessna 152

HAWK XP

Hawk XP

Overall handling characteristics of the Hawk XP show improvement over previous models with the addition of a rounded leading edge on the elevator. The forces required during landing flare are reduced by 35 percent. Balked landing climb capability also has been enhanced and trim requirements during flap setting changes reduced. Part of the basic avionics kit on the Hawk XP includes an improved antiprecipitation static navigation antenna which provides better suppression of undesired signals for more navigation reliability. New options include an avionics cooling system and an intercom system.

Manufacturer and Model	*Seats*	*Powerplant*	*Fuel Capacity (std/opt. gal)*	*Weights Gross/ Empty (lbs)*	*Cruise Speed (mph) 75% @ alt 65% @ alt*	*Optimum Range (m w 45 min. rsv) 75% @ alt 65% @ alt*	*Takeoff/ Landing Distance (ft. over 50′ obst)*	*Rate of Climb (fpm)*	*Service Ceiling*	*Stall Speed (gear/flaps down, mph)*
CESSNA Hawk XP	4	Cont. IO-360-KB 195 hp.	49/66	2558/ 1538	146 @ 6,000′ 137 @ 8,000′	529 @ 6,000′ 587 @ 8,000′	1360/ 1270	870	17,000′	53

Cessna Skyhawk

Skyhawk

The newest Skyhawk features increased useful load, a new engine, improved handling, and a new avionics cooling system. Ramp weight has been increased over previous models. With four 170-pound people on board and nearly 75 pounds for baggage or options, there is enough useful load for 40 gallons of fuel. The engine is the 160 HP Lycoming 0-320-D2J, which provides a maximum speed of 123 knots (141 mph) at 2,700 rpm.

Manufacturer and Model	Seats	Powerplant	Fuel Capacity (std/opt. gal)	Weights Gross/ Empty (lbs)	Cruise Speed (mph) 75% @ alt 65% @ alt	Optimum Range (m w 45 min. rsv) 75% @ alt 65% @ alt	Takeoff/ Landing Distance (ft. over 50′ obst)	Rate of Climb (fpm)	Service Ceiling	Stall Speed (gear/flaps down, mph)
CESSNA Skyhawk	4	Lyc. O-320-H2AD 160 hp.	40/50	2307/ 1403	140 @ 8,000′ 132 @ 8,000′	524 @ 8,000′ 582 @ 8,000′	1390/ 1250	770	14,200′	50

New standard features of the 180 and 185 Skywagons are an antiprecipitation static navigation antenna when factory-installed radios are ordered, an improved avionics cooling system, and an improved fuel selector linkage. Optional equipment includes a new intercom system, an improved approach plate holder, a lightweight headset, and Collins DME and RNAV equipment. There is also an agricultural option to equip both models for aerial application or ag pilot training. The option includes a 151-gallon fan-driven Sorensen spray system with electric spray valve, deflector cable, and windshield, and landing gear wire cutters.

180 and 185 Skywagons

Manufacturer and Model	Seats	Powerplant	Fuel Capacity (std/opt. gal)	Weights Gross/ Empty (lbs)	Cruise Speed (mph) 75% @ alt 65% @ alt	Optimum Range (m w 45 min. rsv) 75% @ alt 65% @ alt	Takeoff/ Landing Distance (ft. over 50′ obst)	Rate of Climb (fpm)	Service Ceiling	Stall Speed (gear/flaps down, mph)
CESSNA 180 Skywagon	6	Cont. O-470-U 230 hp.	84	2810/ 1643	163 @ 8,000′ 153 @ 12,000′	949 @ 8,000′ 990 @ 12,000′	1205/ 1365	1100	17,700′	55
CESSNA 185 Skywagon	6	Cont. 10-520-D 300 hp.	84	3362/ 1688	169 @ 7,000′ 162 @ 10,000′	742 @ 7,000′ 823 @ 10,000′	1430/ 1400	1075	17,900′	56

Cessna 180 and 185 Skywagons

Cessna Turbo Skylane

Skylane and Turbo Skylane

Exterior styling of the newest Skylane includes optional adobe beige as well as standard vestal white base colors in polyurethane paint. New exterior striping and new interior fabrics and colors provide the latest in aircraft styling. The Skylane panel has a new look with standard black or shadow gray panel inserts. A new audio control panel and marker beacon has updated circuitry for more flexibility. Options include a writing table and an AM/FM stereo cassette entertainment center. The Turbo Skylane is powered by the same turbocharged 235 HP Lycoming engine found in the Turbo Skylane RG.

Manufacturer and Model	*Seats*	*Powerplant*	*Fuel Capacity (std/opt. gal)*	*Weights Gross/ Empty (lbs)*	*Cruise Speed (mph) 75% @ alt 65% @ alt*	*Optimum Range (m w 45 min. rsv) 75% @ alt 65% @ alt*	*Takeoff/ Landing Distance (ft. over 50′ obst)*	*Rate of Climb (fpm)*	*Service Ceiling*	*Stall Speed (gear/flaps down, mph)*
CESSNA Skylane	4	Cont. 0-470-U 230 hp.	88	2960/ 1725	166 @ 8,000′ 155 @ 8,000′	967 @ 8,000′ 1065 @ 8,000′	1350/ 1350	1010	16,500′	57
CESSNA Turbo-Skylane	4	Turbo-charged Avco Lycoming O-540-L3C5D 235 bhp at 2400 rpm	92	3100/ 1781	167 @ 10,000′ NA	835 @ 10,000′ NA	965	965	20,000′	56

NA–Not Available

Skylane RG and Turbo Skylane RG

Standard on new models is an improved avionics cooling system enhanced by addition of an electric cooling fan that pulls from cabin air and eliminates the need for an external air scoop. The fan aids cooling during ground operations. A new fiberglass-encapsulated navigation antenna reduces static discharging from the antenna during flight in precipitation, improving VOR and VHF communications performance during flights in precipitation.

Manufacturer and Model	Seats	Powerplant	Fuel Capacity (std/opt. gal)	Weights Gross/ Empty (lbs)	Cruise Speed (mph) 75% @ alt 65% @ alt	Optimum Range (m w 45 min. rsv) 75% @ alt 65% @ alt	Takeoff/ Landing Distance (ft. over 50′ obst)	Rate of Climb (fpm)	Service Ceiling	Stall Speed (gear/flaps down, mph)
CESSNA Skylane RG	4	Lyc. O-540-L3C5D 235 hp.	88	3112/ 1750	179 @ 7,500′ 172 @ 7,500′	972 @ 7,500′ 1080 @ 7,500′	1570/ 1320	1140	14,300′	57
CESSNA Turbo Skylane RG	4	Lyc. O0540-L3C5D 235 hp.	88	3112/ 1791	199 @ 20,000′ 186 @ 20,000′	950 @ 20,000′ 1030 @ 20,000′	1570/ 1320	1040	20,000′	57

Cessna Stationair 6

Stationair 6 and Turbo Stationair 6

The Stationair 6 models can seat six persons or accommodate one-half ton of cargo. Standard features are an improved avionics cooling system, improved cooling vents, a new muffler for the normally aspirated model, an improved fuel selector linkage, antiprecipitation static navigation antenna when factory-installed radios are ordered, and a new battery electrical contactor.

Manufacturer and Model	Seats	Powerplant	Fuel Capacity (std/opt. gal)	Weights Gross/ Empty (lbs)	Cruise Speed (mph) 75% @ alt 65% @ alt	Optimum Range (m w 45 min. rsv) 75% @ alt 65% @ alt	Takeoff/ Landing Distance (ft. over 50′ obst)	Rate of Climb (fpm)	Service Ceiling	Stall Speed (gear/flaps down, mph)
CESSNA Turbo Stationair 6	6	Cont. TSIO-520-M 310 hp.	88	3616/ 2003	169 @ 22,000′ 157 @ 22,000′	771 @ 22,000′ 829 @ 22,000′	1640/ 1395	1010	27,000′	62
CESSNA Staitionair 6	6	Cont. 1O-520-F 300 hp.	88	3612/ 1927	153 @ 6,500′ 149 @ 10,000′	783 @ 6,500′ 875 @ 10,000′	1780/ 1395	920	14,800	62

Centurion, Turbo Centurion, and Pressurized Centurion

New standard features for the Centurion and Turbo Centurion include an improved avionics cooling system and new air outlets for the overhead ventilation system. Avionics cooling has been improved with the addition of an electric cooling fan that pulls in cabin air and eliminates the need for an external air scoop. The fan aids cooling during ground operations.

Cessna Centurian

Manufacturer and Model	Seats	Powerplant	Fuel Capacity (std/opt. gal)	Weights Gross/ Empty (lbs)	Cruise Speed (mph) 75% @ alt 65% @ alt	Optimum Range (m w 45 min. rsv) 75% @ alt 65% @ alt	Takeoff/ Landing Distance (ft. over 50′ obst)	Rate of Climb (fpm)	Service Ceiling	Stall Speed (gear/flaps down, mph)
CESSNA Centurion	6	Cont. IO-520-L 300 hp.	89	3812/ 2133	197 @ 6,500′ 192 @ 10,000′	926 @ 6,500′ 1047 @ 10,000′	2030/ 1500	950	17,300′	64
CESSNA Turbo Centurion	6	Cont. TS10-520-R 310 hp.	89	4016/ 2221	222 @ 22,000′ 208 @ 24,000′	909 @ 22,000′ 978 @ 24,000′	2160/ 1500	930	27,000′	66
CESSNA Pressurized Centurion	6	Cont. TSTO-520-P 310 hp.	89	4016/ 2340	222 @ 22,000′ 208 @ 22,000′	840 @ 22,000′ 897 @ 22,000′	2160/ 1500	930	23,000′	66

Cessna Pressurized Centurian

Cessna Staionair 8

Stationair 8 and Turbo Stationair 8

New standard features include an improved battery electrical contactor, new wing root vents, an antiprecipitation static navigation antenna, new avionics cooling system, and a more effective muffler for the normally aspirated Stationair 8. Optional features include a high-capacity alternator, a cabin intercom system, an improved approach plate holder, refueling steps and handles, and additional avionics options.

Manufacturer and Model	Seats	Powerplant	Fuel Capacity (std/opt. gal)	Weights Gross/ Empty (lbs)	Cruise Speed (mph) 75% @ alt 65% @ alt	Optimum Range (m w 45 min. rsv) 75% @ alt 65% @ alt	Takeoff/ Landing Distance (ft. over 50′ obst)	Rate of Climb (fpm)	Service Ceiling	Stall Speed (gear/flaps down, mph)
CESSNA Stationair 8	8	Cont. IO-520-F 300 hp.	54/73	3812/ 2105	164 @ 6,500′ 158 @ 10,000′	402 @ 6,500′ 449 @ 10,000′	1970/ 1500	810	13,300′	66
CESSNA Turbo Stationair 8	8	Cont. TS10-520-M 310 hp.	54/73	3816/ 2183	180 @ 20,000′ 166 @ 20,000′	362 @ 20,000′ 402 @ 20,000′	1860/ 1500	885	26,000′	66

Cutlass RG

Takeoff weight of the Cutlass RG is 2,650 pounds and useful load is 1,103 pounds. Two full-size doors plus a baggage door make loading the airplane easy. The baggage area is certified for 200 pounds. Gear operating speed is 140 knots, and the first ten degrees of flaps can be lowered at 130 knots, allowing rapid descents. That, in turn, allows an ease of operation in flying a variety of approach procedures in heavy traffic. New standard features are an avionics cooling system, an oil pressure gauge, and an improved navigation antenna.

Cessna Cutlass RG

Manufacturer and Model	*Seats*	*Powerplant*	*Fuel Capacity (std/opt. gal)*	*Weights Gross/ Empty (lbs)*	*Cruise Speed (mph) 75% @ alt 65% @ alt*	*Optimum Range (m w 45 min. rsv) 75% @ alt 65% @ alt*	*Takeoff/ Landing Distance (ft. over 50′ obst)*	*Rate of Climb (fpm)*	*Service Ceiling*	*Stall Speed (gear/flaps down, mph)*
CESSNA Cutlass RG	4	Lyc. O-360-F1A6 180 hp.	62	2658/ 1558	161 @ 9,000′ 149 @ 9,000′	828 @ 9,000′ 898 @ 9,000′	1775/ 1340	800	16,800′	57

LAKE AIRCRAFT

Lake Buccaneer

The Lake Buccaneer is a four-place production amphibian aircraft. It can take off and land from water or land, and it can push four people and baggage along at 150 mph. The plane is alodined and zinc-chromated inside and out for saltwater operation.

Manufacturer and Model	Seats	Powerplant	Fuel Capacity (std/opt. gal)	Weights Gross/ Empty (lbs)	Cruise Speed (mph) 75% @ alt 65% @ alt	Optimum Range (m w 45 min. rsv) 75% @ alt 65% @ alt	Takeoff/ Landing Distance (ft. over 50′ obst)	Rate of Climb (fpm)	Service Ceiling	Stall Speed (gear/flaps down, mph)
LAKE AIRCRAFT Lake Buccaneer	4	Lyc. fuel-injected, 200 hp.	54	2690/ 1555	150	825/NA	600/475 (Land), 1100/600 (water)	1200	NA	45

NA–Not Available

Lake Buccaneer

Lake Buccaneer

MAULE

Lunar Rockets

Maule produces several versions of the M-5. Equipped with skis or floats, the Maule M-5 Lunar Rocket flies in and out of the most inaccessible regions. The newest M-5 is the M-6 Super Stol Rocket, which has a new wing design.

Manufacturer and Model	*Seats*	*Powerplant*	*Fuel Capacity (std/opt. gal)*	*Weights Gross/ Empty (lbs)*	*Cruise Speed (mph) 75% @ alt 65% @ alt*	*Optimum Range (m w 45 min. rsv) 75% @ alt 65% @ alt*	*Takeoff/ Landing Distance (ft. over 50′ obst)*	*Rate of Climb (fpm)*	*Service Ceiling*	*Stall Speed (gear/flaps down, mph)*
MAULE M-5-180C	4	Lyc. O-360- C1F 180 hp.	40/63	2900/ 1300	156 @ 7,500′ 149 @ 7,500′	700 @ 7,500′ 1050 @ 7,500′	600/ 600	900	15,000′	38 STOL aircraft
MAULE M-5-210C	4	Cont. 1O-360-D 210 hp.	40/63	2500/ 1350	158 @ 8,500′ NA	600 @ 8,500′ 950 @ 8,500′	600/ 600	1250	18,000′	38 STOL aircraft
MAULE M-5-210TC	4	Lyc. TO-360-F1A6D 210 hp.	40/63	2500/ 1400	196 @ 7,500′ 180 @ 7,500′	550 @ 7,500′ 865 @ 7,500′	600/ 600	1250	20,000′	38 STOL aircraft
MAULE M-5-235C	4	Lyc. O-540-J1A5D 235 hp.	40/63	2500/ 1400	172 @ 7,500′ 163 @ 7,500′	550 @ 7,500′ 865 @ 7,500′	600/ 600	1350	20,000′	38 STOL aircraft
MAULE M-6	4	NA	69	2500/ 1450	NA	NA	NA	NA	NA	26 STOL aircraft

NA–Not Available

Maule M-5 180C Lunar Rocket

MOONEY

Mooney 201

201 A 200 HP plane with a top speed over 200 mph, the 201 handles four adults, has a cabin measuring 43.5 inches elbow-to-elbow, and offers a choice of nine autopilot systems and 14 different avionic packages.

Manufacturer and Model	*Seats*	*Powerplant*	*Fuel Capacity (std/opt. gal)*	*Weights Gross/ Empty (lbs)*	*Cruise Speed (mph) 75% @ alt 65% @ alt*	*Optimum Range (m w 45 min. rsv) 75% @ alt 65% @ alt*	*Takeoff/ Landing Distance (ft. over 50′ obst)*	*Rate of Climb (fpm)*	*Service Ceiling*	*Stall Speed (gear/flaps down, mph)*
MOONEY MsOJ 201	4	Lyc. IO-360-A3B6 200 hp.	64	2740/ 1640	194 @ 8,000′ 184 @ 8,000′	925 @ 6,000′ 1063 @ 6,000′	1517/ 1610	1030	18,800′	61

The Turbo 231 has an engine rated at 210 HP and a top speed of 231 mph, room for four adults, 43.5 inches elbow-to-elbow, and contour seats in a wide choice of genuine leather fabrics or leatherlike vinyls. The 231 comes with a wide choice of avionics equipment.

231

Manufacturer and Model	*Seats*	*Powerplant*	*Fuel Capacity (std/opt. gal)*	*Weights Gross/ Empty (lbs)*	*Cruise Speed (mph) 75% @ alt 65% @ alt*	*Optimum Range (m w 45 min. rsv) 75% @ alt 65% @ alt*	*Takeoff/ Landing Distance (ft. over 50′ obst)*	*Rate of Climb (fpm)*	*Service Ceiling*	*Stall Speed (gear/flaps down, mph)*
MOONEY M20K 231	4	Cont. TSIO-360-GB 210 hp.	72	2900/ 1800	209 @ 18,000′ 206 @ 24,000′	1093 @ 18,000′ 1180 @ 8,000′	2060/ 2280	1080	24,000′	65

Mooney 231

PIPER

Piper Tomahawk

Tomahawk

Design features include rugged spring landing gear, ten feet wide for maximum stability; modern low wing for inflight visibility, ground handling safety and landing ease, secure tie-down in gusty winds; cabin entrance doors on each side with wing walks and optional steps; and three-position fuel selector valve in center of panel, in line of sight.

Manufacturer and Model	Seats	Powerplant	Fuel Capacity (std/opt. gal)	Weights Gross/ Empty (lbs)	Cruise Speed (mph) 75% @ alt 65% @ alt	Optimum Range (m w 45 min. rsv) 75% @ alt 65% @ alt	Takeoff/ Landing Distance (ft. over 50′ obst)	Rate of Climb (fpm)	Service Ceiling	Stall Speed (gear/flaps down, mph)
PIPER Tomahawk	2	Lyc. O-235-L2C 112 hp.	32	1670/ 1108	124 @ 7,100′ 115 @ 10,500′	520 @ 7,100′ 538 @ 10,500′	1460/ 1544	718	13,000′	56

A Lycoming O-330 engine powers the Super Cub, which includes such features as high life metal wing flaps, upholstered front and rear seats, a baggage compartment of 18 cubic feet, cabin heater, and aerodynamically balanced rudder and elevators.

Super Cub

Piper Super Club

Manufacturer and Model	Seats	Powerplant	Fuel Capacity (std/opt. gal)	Weights Gross/ Empty (lbs)	Cruise Speed (mph) 75% @ alt 65% @ alt	Optimum Range (m w 45 min. rsv) 75% @ alt 65% @ alt	Takeoff/ Landing Distance (ft. over 50′ obst)	Rate of Climb (fpm)	Service Ceiling	Stall Speed (gear/flaps down, mph)
PIPER Super Cub	2	Lyc. O-320 150 hp.	36	1750/ 983	115 @ 5,000′ NA	460 @ 5,000′ NA	500/ 885	960	19,000′	42

NA–Not Available

Piper Warrior II

Warrior II

With a 160 Lycoming engine, the Warrior II can travel to 127 knots (146 mph) cruise speed. The airplane can carry a thousand pounds of useful load. Inside the plane are 24 cubic feet that can transport up to 200 pounds of baggage. Individual ventilation systems offer fingertip control in an overhead console to each passenger on board the plane.

Manufacturer and Model	Seats	Powerplant	Fuel Capacity (std/opt. gal)	Weights Gross/ Empty (lbs)	Cruise Speed (mph) 75% @ alt 65% @ alt	Optimum Range (m w 45 min. rsv) 75% @ alt 65% @ alt	Takeoff/ Landing Distance (ft. over 50′ obst)	Rate of Climb (fpm)	Service Ceiling	Stall Speed (gear/flaps down, mph)
PIPER Warrior II	4	Lyc. O-320-D3G 160 hp.	50	2325/ 1340	146 @ 9,000′ 136 @ 12,500′	679 @ 9,000′ 729 @ 12,500′	1490/ 1115	710	14,000′	57

Piper Archer II

Archer II

The semitapered wing configuration of the Archer is teamed with a 180 HP four-cylinder Lycoming engine to deliver performance of 129 kts. cruise speed without sacrificing energy or cost efficiency. The cabin can seat four with individually controlled outlets for heating, defrosting, and ventilation systems. Deluxe interior features include crushed velour, deep pile carpeting, and attractive cushions.

Manufacturer and Model	Seats	Powerplant	Fuel Capacity (std/opt. gal)	Weights Gross/ Empty (lbs)	Cruise Speed (mph) 75% @ alt 65% @ alt	Optimum Range (m w 45 min. rsv) 75% @ alt 65% @ alt	Takeoff/ Landing Distance (ft. over 50′ obst)	Rate of Climb (fpm)	Service Ceiling	Stall Speed (gear/flaps down, mph)
PIPER Archer II	4	Lyc. O-360-A4M 180 hp.	50	2550/ 1418	148 @ 8,000′ 144 @ 12,000′	690 @ 8,000′ 742 @ 12,000′	1625/ 1400	735	15,000′	61

Dakota

The Dakota cruises at 144 knots (166 mph) and has a rate of climb of 1,110 fpm. Deluxe interior features include crushed velour, deep pile carpeting, and attractive cushions. The new heavy-duty brake option improves the plane's performance. The wide passenger and cargo doors make it easy to load and unload the plane of people and goods.

Piper Dakota

Manufacturer and Model	*Seats*	*Powerplant*	*Fuel Capacity (std/opt. gal)*	*Weights Gross/ Empty (lbs)*	*Cruise Speed (mph) 75% @ alt 65% @ alt*	*Optimum Range (m w 45 min. rsv) 75% @ alt 65% @ alt*	*Takeoff/ Landing Distance (ft. over 50' obst)*	*Rate of Climb (fpm)*	*Service Ceiling*	*Stall Speed (gear/flaps down, mph)*
PIPER Dakota	4	Lyc. O-540-J3A5D 235 hp.	77	3000/ 1634	165 @ 9,100' 158 @ 12,200'	817 @ 8,500' 880 @ 11,400'	1216/ 1410	1110	17,500'	64
PIPER Turbo Dakota	4	Cont. TSIO-360-FB 200 hp.	77	2900/ 1579	165 @ 20,000' 158 @ 20,000'	817 @ 20,000' 880 @ 20,000'	1402/ 1697	902	20,000'	66

Arrow IV and Turbo Arrow IV

With its 200 HP Lycoming engine, the normally aspirated Arrow cruises at 143 knots (almost 165 mph), and it carries over a half-ton of useful load. The turbocharged 200 HP Continental engine on the Turbo Arrow gives full-rated takeoff power from airports at high elevations.

Manufacturer and Model	Seats	Powerplant	Fuel Capacity (std/opt. gal)	Weights Gross/ Empty (lbs)	Cruise Speed (mph) 75% @ alt 65% @ alt	Optimum Range (m w 45 min. rsv) 75% @ alt 65% @ alt	Takeoff/ Landing Distance (ft. over 50′ obst)	Rate of Climb (fpm)	Service Ceiling	Stall Speed (gear/flaps down, mph)
PIPER Arrow IV	4	Lyc. IO-360-C1C6 200 hp.	77	2750/ 1641	164 @ NA 158 @ NA	933 @ NA 973 @ NA	1600/ 1525	831	17,000′	63
PIPER Turbo Arrow IV	4	Cont. TSIO-360-F 200 hp.	77	2900/ 1690	198 @ NA 192 @ NA	898 @ NA 956 @ NA	1620/ 1555	940	20,000′	66

NA–Not available

Piper Turbo Arrow IV

Saratoga and Turbo Saratoga

Baggage can be stored easily in the Saratoga, in both fore and aft compartments and, if more space is needed, seats can be slided out so that the entire interior becomes a cavernous cargo hold.

Piper Saratoga

Manufacturer and Model	Seats	Powerplant	Fuel Capacity (std/opt. gal)	Weights Gross/ Empty (lbs)	Cruise Speed (mph) 75% @ alt 65% @ alt	Optimum Range (m w 45 min. rsv) 75% @ alt 65% @ alt	Takeoff/ Landing Distance (ft. over 50′ obst)	Rate of Climb (fpm)	Service Ceiling	Stall Speed (gear/flaps down, mph)
PIPER Saratoga	6	Lyc. IO-540-K1G5D 300 hp.	107	3615/ 1920	192 @ 8,000′ 168 @ 10,000′	947 @ 8,000′ 1048 @ 10,000′	1573/ 1530	990	14,100′	69
PIPER Turbo Saratoga	6	Lyc. TIO-540-SIAD 300 hp.	107	2000/ 1617	189 @ 20,000′ 177 @ 20,000′	898 @ 20,000′ 972 @ 20,000′	1420/ 1700	1075	20,000′	69

Piper Turbo Saratoga SP

Saratoga SP and Turbo Saratoga SP

The Saratoga SP has a 300 HP Lycoming engine. Fuel tanks have a 102-gallon capacity, allowing the owner to fly 1,132 statute miles and still maintain a 45-minute fuel reserve. Cruise speed is more than 180 mph. The Turbo Saratoga SP is equipped with an optional three-bladed propeller, and has a top speed of 225 mph.

Manufacturer and Model	Seats	Powerplant	Fuel Capacity (std/opt. gal)	Weights Gross/ Empty (lbs)	Cruise Speed (mph) 75% @ alt 65% @ alt	Optimum Range (m w 45 min. rsv) 75% @ alt 65% @ alt	Takeoff/ Landing Distance (ft. over 50′ obst)	Rate of Climb (fpm)	Service Ceiling	Stall Speed (gear/flaps down, mph)
PIPER Saratoga SP	6	Lyc. IO-540-K1G5D 300 hp.	107	3600/ 1986	182 @ NA 176 @ NA	898 @ NA 949 @ NA	1573/ 1612	1010	16,700′	68
PIPER Turbo Saratoga SP	6	Lyc. TIO-540-S1AD 300 hp.	107	3600/ 2073	203 @ 20,000′ 190 @ 20,000′	971 @ 20,000′ 1059 @ 20,000′	1420/ 1640	1120	20,000′	69

NA–Not Available

TAYLORCRAFT

Taylorcraft F21

F21

The F21 includes such standard equipment as hydraulic toe brakes, 12-gallon fuselage tank, two all-metal doors with sliding windows, and an 85-pound baggage compartment with side windows. Optional equipment items are a landing light in the left wing, a cabin light, emergency locator transmitter, headset, headphone, and speakers.

Manufacturer and Model	Seats	Powerplant	Fuel Capacity (std/opt. gal)	Weights Gross/ Empty (lbs)	Cruise Speed (mph) 75% @ alt 65% @ alt	Optimum Range (m w 45 min. rsv) 75% @ alt 65% @ alt	Takeoff/ Landing Distance (ft. over 50′ obst)	Rate of Climb (fpm)	Service Ceiling	Stall Speed (gear/flaps down, mph)
TAYLOR CRAFT F21		Lyc. O.235 L2	24	1500/ 990	122 @ 8,000′ NA	400 @ 8,000′ NA	350/ 350	875	18,000′	43

NA/-Not Available

VARGA AIRCRAFT

2150A

The 2150A model by Varga is powered by a Lycoming engine, O-320-A2C. Its gross weight in pounds is 1,817 and its empty weight in pounds is 1,125.

Varga Aircraft 2150A

Manufacturer and Model	Seats	Powerplant	Fuel Capacity (std/opt. gal)	Weights Gross/ Empty (lbs)	Cruise Speed (mph) 75% @ alt 65% @ alt	Optimum Range (m w 45 min. rsv) 75% @ alt 65% @ alt	Takeoff/ Landing Distance (ft. over 50′ obst)	Rate of Climb (fpm)	Service Ceiling	Stall Speed (gear/flaps down, mph)
VARGA Kachina 2150A	2	Lyc. O-320-A2C 150 hp.	35	1817/ 1125	134 @ 1,000′ 126 @ 1,000′	425 @ 1,000′ 526 @ 1,000′	440/ 450	1450	22,000′	52

Varga 2180

2180

The 2180 is also powered by the Lycoming O-320A engine, the same one in the 2150A Varga. All Varga models can carry up to 50 pounds and have a fuel capacity in gallons of 35.

Manufacturer and Model	Seats	Powerplant	Fuel Capacity (std/opt. gal)	Weights Gross/ Empty (lbs)	Cruise Speed (mph) 75% @ alt 65% @ alt	Optimum Range (m w 45 min. rsv) 75% @ alt 65% @ alt	Takeoff/ Landing Distance (ft. over 50′ obst)	Rate of Climb (fpm)	Service Ceiling	Stall Speed (gear/flaps down, mph)
VARGA 2180		Lyc. O-320A	35	1870/ 1175	133 @ 10,000′ 125 @ 10,000′	451 @ 10,000′ 472 @ 10,000′	1310/ 1425	NA	22,000′	52

Multi-Engine Aircraft

Beechcraft	9709 East Central Wichita, Kansas 67201	Duchess Baron B55 Baron E55 Baron 58 Baron 58TC Baron 58P Duke B60
Cessna	P.O Box 1521 Wichita, Kansas 67201	402, Businessliner, and Utiliner 340 Chancellor Crusader 303 Titan (Courier, Ambassador, and Freighter) 421 Golden Eagle
Pilatus Britten-Norman	Bembridge Airport Bembridge, Isle of Wight PO35 5PR England U.S. Representative: Jonas Aircraft 120 Wall Street New York, New York 10005	Islander BN2B Islander BN2T Trislander
Piper	Lock Haven, Pennsylvania 17745	Seminole and Turbo Seminole Seneca II Aerostar (601B and 601P) Navajo and Navajo C/R Chieftain
Wing	2925 Columbia Street Torrance, California 90503	D-1 Derringer

BEECHCRAFT

Duchess

Two Lycoming four-cylinder O-360 engines rated at 180 HP power. The Duchess includes such standard cabin equipment as left and right cabin entrance doors, left side baggage door, super soundproofing, two-place rear bench seat, windshield defroster, six adjustable fresh air vents, two heat and fresh air vents, map stowage, and rear accessory storage shelf.

Beechcraft Duchess

Manufacturer and Model	Seats	Powerplants	Fuel Capacity (std/opt. gal)	Weights Gross/ Empty (lbs)	Cruise Speed (mph) 75% @ alt 65% @ alt	Optimum Range (m w 45 min. rsv) 75% @ alt 65% @ alt	Takeoff/ Landing Distance (ft. over 50' obst)	Rate of Climb Engine out ROC (fpm)	Service Ceiling	Stall Speed (gear/flaps down, mph)
BEECH Duchess 76	4	2 Lyc. O-360-A1G6D 180 hp ea.	100	3900/ 2460	188 @ 8,000′ 179 @ 8,000′	751 @ 8,000′ 818 @ 10,000′	2119/ 1181	1248/ 235	19,650′	69

The Baron B55 includes an avionics package with Com Transceiver, Nav Receiver, microphone, headset, cabin speaker, and Nav/GS and Com antennas. The two engines are Continental six-cylinder 10-470 fuel-injected models rated at 260 HP.

Baron B55

Beechcraft Baron B55

Manufacturer and Model	Seats	Powerplants	Fuel Capacity (std/opt. gal)	Weights Gross/ Empty (lbs)	Cruise Speed (mph) 75% @ alt 65% @ alt	Optimum Range (m w 45 min. rsv) 75% @ alt 65% @ alt	Takeoff/ Landing Distance (ft. over 50′ obst)	Rate of Climb Engine out ROC (fpm)	Service Ceiling	Stall Speed (gear/flaps down, mph)
BEECH Baron B55	4-6	2 Cont. IO-470-L 260 hp ea.	100/136	5100/ 3233	215 @ 7,000′ 207 @ 8,000′	949 @ 7,000′ 1044 @ 10,000′	2154/ 2148	1693/ 397	19,300′	84

Baron E55

Not only does the E55 include a standard King avionics package, but also such important equipment as three-control system with adjustable pilot rudder pedals, dual electric tachometer, airspeed indicator, sensitive altimeter, instrument panel flood lighting in glareshield, two landing lights, quick removable gas tank caps, and emergency locator transmitter.

Manufacturer and Model	Seats	Powerplants	Fuel Capacity (std/opt. gal)	Weights Gross/ Empty (lbs)	Cruise Speed (mph) 75% @ alt 65% @ alt	Optimum Range (m w 45 min. rsv) 75% @ alt 65% @ alt	Takeoff/ Landing Distance (ft. over 50′ obst)	Rate of Climb Engine out ROC (fpm)	Service Ceiling	Stall Speed (gear/flaps down, mph)
BEECH Baron E55	4-6	2 Cont. IO-520-CB 285 hp ea.	100/136/ 166	5300/ 3286	229 @ 7,000′ 218 @ 8,000′	1102 @ 7,000′ 1188 @ 10,000′	2050/ 2202	1682/ 388	19,100′	84

Beechcraft Baron E55

Baron 58

Baggage capacity is unusually spacious in the Baron 58 with 300-pound capacity in the nose compartment, 200-pound capacity in the center cabin between spars, 400 pounds in the rear compartment, and 120 pounds in the extended rear compartment. To facilitate night flights, there are two landing lights, position lights, landing gear position lights, map light, utility door-ajar light, and entrance door light.

Beechcraft Baron 58

Manufacturer and Model	Seats	Powerplants	Fuel Capacity (std/opt. gal)	Weights Gross/ Empty (lbs)	Cruise Speed (mph) 75% @ alt 65% @ alt	Optimum Range (m w 45 min. rsv) 75% @ alt 65% @ alt	Takeoff/ Landing Distance (ft. over 50′ obst)	Rate of Climb Engine out ROC (fpm)	Service Ceiling	Stall Speed (gear/flaps down, mph)
BEECH Baron 58	4-6	2 Cont. IO-520-CB 285 hp ea.	136/166/ 194	5400/ 3363	229 @ 7,000′ 218 @ 8,000′	1312 @ 7,000′ 1408 @ 10,000′	2101/ 2498	1660/ 390	18,600′	85

Baron 58TC

Powered with two Continental six-cylinder TSIO-520-WB turbocharged engines rated at 325 HP each, the Baron 58TC includes a standard King avionics package. Baggage capacity exceeds 1,300 pounds in four separate compartments. Service equipment on board: tow bar, service information kit, pitot tube cover, operating handbook and FAA-approved airplane flight manual, engine log books, airplane log book, control lock assembly, flight bag, sump drain wrench.

Beechcraft Baron 58TC

Manufacturer and Model	Seats	Powerplants	Fuel Capacity (std/opt. gal)	Weights Gross/ Empty (lbs)	Cruise Speed (mph) 75% @ alt 65% @ alt	Optimum Range (m w 45 min. rsv) 75% @ alt 65% @ alt	Takeoff/ Landing Distance (ft. over 50′ obst)	Rate of Climb Engine out ROC (fpm)	Service Ceiling	Stall Speed (gear/flaps down, mph)
BEECH Baron 58TC	4-6	2 Cont. TSIO-520-WB 325 hp ea.	166/190	6200/ 3793	272 @ 25,000′ 255 @ 25,000′	1173 @ 25,000′ 1258 @ 25,000′	2643/ 2427	1475/ 270	25,000′	90

The pressurized Baron 58P has a maximum speed of 300 mph, powered by two Continental six-cylinder TSIO-520-WB turbocharged engines rated at 325 HP each. Cabin pressurization equipment includes cabin altitude selector, combined cabin altitude and differential gauge, cabin rate of climb indicator, cabin outflow control valve, cabin pressurization safety valve, inflating front door seal with electric pump, and reserve tank.

Baron 58P

Manufacturer and Model	*Seats*	*Powerplants*	*Fuel Capacity (std/opt. gal)*	*Weights Gross/ Empty (lbs)*	*Cruise Speed (mph) 75% @ alt 65% @ alt*	*Optimum Range (m w 45 min. rsv) 75% @ alt 65% @ alt*	*Takeoff/ Landing Distance (ft. over 50′ obst)*	*Rate of Climb Engine out ROC (fpm)*	*Service Ceiling*	*Stall Speed (gear/flaps down, mph)*
BEECH Baron 58P	4-6	2 Cont. TSIO-520-WB 325 hp ea.	166/190	6200/ 4020	272 @ 25,000′ 255 @ 25,000′	1172 @ 25,000′ 1257 @ 25,000′	2643/ 2427	1475/ 270	25,000′	90

1981 Beechcraft Baron 58P

Duke B60

A four-to-six-place, twin-engine, turbocharged, pressurized aircraft, the Duke B60 features standard equipment of avionics, engine, and flight instruments. Amenities that are built into the plane include inflight cabin storage pockets, ashtray for each seat, cigarette lighter in passenger compartment plus one for pilot and copilot, coat hanger rod, arm rests, four reclining track-mounted seats, headrest and lap belt for all seats, and baggage strap installation for rear of cabin.

Manufacturer and Model	Seats	Powerplants	Fuel Capacity (std/opt. gal)	Weights Gross/ Empty (lbs)	Cruise Speed (mph) 75% @ alt 65% @ alt	Optimum Range (m w 45 min. rsv) 75% @ alt 65% @ alt	Takeoff/ Landing Distance (ft. over 50′ obst)	Rate of Climb Engine out ROC (fpm)	Service Ceiling	Stall Speed (gear/flaps down, mph)
BEECH Duke B60	4-6	2 Lyc. TIO-541-E1C4 380 hp ea.	142/202/ 232	6775/ 4406	276 @ 26,000′ 238 @ 18,000′	1225 @ 26,000′ 1344 @ 25,000′ (63%)	2626/ 3065	1601/ 307	30,000′	84

Beechcraft Duke B60

Cessna 402

402, Businessliner, and Utiliner Models

The 402 is available in Businessliner and Utiliner versions. The Businessliner can seat up to eight people, the Utiliner seats up to 10 people. An option on the Utiliner model is a passenger instruction sign that illuminates "fasten seat belt" and "no smoking" in international symbols. The signs are recessed into the right-hand side of the headliner aft of the cabin divider and activated from the instrument panel. The interior fabric selection for both models includes two new textured fabrics, two new patterns and one solid color fabric. These can be matched with a wide selection of vinyls and leathers to suit individual style and taste.

Manufacturer and Model	Seats	Powerplants	Fuel Capacity (std/opt. gal)	Weights Gross/ Empty (lbs)	Cruise Speed (mph) 75% @ alt 65% @ alt	Optimum Range (m w 45 min. rsv) 75% @ alt 65% @ alt	Takeoff/ Landing Distance (ft. over 50′ obst)	Rate of Climb Engine out ROC (fpm)	Service Ceiling	Stall Speed (gear/flaps down, mph)
CESSNA 402	6-10	2 Cont. TSIO-520-VB 325 hp ea.	213.4	6885/ NA	243 @ 20,000′ 191 @ 20,000′	1132 @ 20,000′ 1420 @ 20,000′	2195/ 2485	1450/ 301	26,900′	78
CESSNA 402C Utiliner	6-10	2 Cont. TSIO-520-VB 325 hp ea.	206	6885/ 4102	241 @ 20,000′ 231 @ 20,000′	1129 @ 20,000′ 1198 @ 20,000′	2195/ 2485	1450/ 301	26,900′	78
CESSNA 402C Businessliner	6-10	2 Cont. TSIO-520-VB 325 hp ea.	206	6885/ 4074	241 @ 20,000′ 231 @ 20,000′	1129 @ 20,000′ 1198 @ 20,000′	2195/ 2485	1450/ 301	26,900′	78

NA–Not Available

Cessna 340

340

The 340 has a comfortable six-place pressurized cabin that can be personalized with a wide variety of interior appointments. An air stair door and wide aisle make passenger loading easy. Baggage capacity totals 34 cubic feet and 590 pounds in wing lockers and nose compartment, and 18.5 cubic feet and 340 pounds in the aft cabin. With optional equipment, the 340 is approved for flight into icing conditions. Other popular options include a refreshment center, club seating, eight-track stereo, vertically adjustable seats for pilot and co-pilot, and flight deck divider curtains.

Manufacturer and Model	Seats	Powerplants	Fuel Capacity (std/opt. gal)	Weights Gross/ Empty (lbs)	Cruise Speed (mph) 75% @ alt 65% @ alt	Optimum Range (m w 45 min. rsv) 75% @ alt 65% @ alt	Takeoff/ Landing Distance (ft. over 50′ obst)	Rate of Climb Engine out ROC (fpm)	Service Ceiling	Stall Speed (gear/flaps down, mph)
CESSNA 340	6	2 Cont. TSIO-520-NB 310 hp ea.	102/207	6025/ 3911	263 @ 25,000′ 245 @ 25,000′	500 @ 25,000′ 535 @ 25,000′	2175/ 1850	1650/ 315	29,800′	81

Chancellor

Six-place seating is standard in the Chancellor; eight-place seating is optional. A full range of executive options inside the cabin include stereo, refreshment center, air conditioning, club seating, writing tables, flight deck divider, and flushing toilet. A new selection of five fabrics is available for interior styling: two textured fabrics, two patterns and one solid color fabric. These can be color coordinated with a wide selection of vinyls and leathers.

Cessna Chancellor

Manufacturer and Model	*Seats*	*Powerplants*	*Fuel Capacity (std/opt. gal)*	*Weights Gross/ Empty (lbs)*	*Cruise Speed (mph) 75% @ alt 65% @ alt*	*Optimum Range (m w 45 min. rsv) 75% @ alt 65% @ alt*	*Takeoff/ Landing Distance (ft. over 50′ obst)*	*Rate of Climb Engine out ROC (fpm)*	*Service Ceiling*	*Stall Speed (gear/flaps down, mph)*
CESSNA Chancellor	6-8	2 Cont. TSIO-520-NB 310 hp ea.	206	6785/ 4356	257 @ 25,000′ 239 @ 25,000′	1265 @ 25,000′ 1369 @ 25,000′	2595/ 2393	1580/ 290	30,800′	82

Crusader 303

Crusader 303

The Crusader 303 has not yet been certified, although it is expected to be certified in the immediate future.

Manufacturer and Model	Seats	Powerplants	Fuel Capacity (std/opt. gal)	Weights Gross/ Empty (lbs)	Cruise Speed (mph) 75% @ alt 65% @ alt	Optimum Range (m w 45 min. rsv) 75% @ alt 65% @ alt	Takeoff/ Landing Distance (ft. over 50′ obst)	Rate of Climb Engine out ROC (fpm)	Service Ceiling	Stall Speed (gear/flaps down, mph)
CESSNA Crusader 303	6	Teledyne Continental TSIO-520-AE, 250 hp.	155	5175/ 3305	NA	NA	2100/ 1535	1570 NA	25,000′	72

NA–Not Available

Titan (Courier, Ambassador, and Freighter Models)

The Courier version can fly up to 965 nautical (1,110 statute) miles with 11 people and baggage. With quick seat removal, it can work to haul cargo. The Ambassador can seat up to 10 people. The Freighter has a 235-cubic-foot cargo capacity. Standard equipment includes the double cargo door and a crew door to allow the pilot to board after the plane is loaded.

Manufacturer and Model	Seats	Powerplants	Fuel Capacity (std/opt. gal)	Weights Gross/ Empty (lbs)	Cruise Speed (mph) 75% @ alt 65% @ alt	Optimum Range (m w 45 min. rsv) 75% @ alt 65% @ alt	Takeoff/ Landing Distance (ft. over 50′ obst)	Rate of Climb Engine out ROC (fpm)	Service Ceiling	Stall Speed (gear/flaps down, mph)
CESSNA Titan Courier	10-11	2 Cont. GTSIO-520-M 375 hp ea.	344	8450/ 4844	250 @ 20,000′ 230 @ 20,000′	1145 @ 20,000′ 1231 @ 20,000′	2367/ 2130	1575/ 230	26,000′	80
CESSNA Titan Ambassador	6-10	2 Cont. GTSIO-520-M 375 hp ea.	344	8450/ 4816	250 @ 20,000′ 230 @ 20,000′	1146 @ 20,000′ 1232 @ 20,000′	2367/ 2130	1575/ 230	26,000′	80
CESSNA Titan Freighter	2	2 Cont. GTSIO-520-M 375 hp ea.	344	8450/ 4685	250 @ 20,000′ 250 @ 20,000′	1146 @ 20,000′ 1232 @ 20,000′	2367/ 2130	1575/ 230	26,000′	80

Cessna Titan

421 Golden Eagle

Improvements to the fuel injection system include a new engine-driven fuel pump that maintains pressure with a regulator referenced to the engine turbocharging system. An engine injector system provides precise fuel flow for smooth operation at all altitudes and a priming system independent of the fuel injection system enhances engine starting during temperature extremes. The 421 also features a pneumatic lower cabin door extender as standard equipment that lowers the cabin door gently and smoothly. Frost panes on cockpit side windows eliminate frost and result in better visibility. Polished propeller spinners are available as a styling option.

Manufacturer and Model	Seats	Powerplants	Fuel Capacity (std/opt. gal)	Weights Gross/ Empty (lbs)	Cruise Speed (mph) 75% @ alt 65% @ alt	Optimum Range (m w 45 min. rsv) 75% @ alt 65% @ alt	Takeoff/ Landing Distance (ft. over 50′ obst)	Rate of Climb Engine out ROC (fpm)	Service Ceiling	Stall Speed (gear/flaps down, mph)
CESSNA 421 Golden Eagle	6-8	2 Cont. TSIO-520-L 375 hp ea.	206/262	7500/ 4623	277 @ 25,000′ 261 @ 25,000′	1018 @ 25,000′ 1093 @ 25,000′	2323/ 2293	1940/ 350	30,200′	85

Cessna 421 Golden Eagle

Pilatus Britten-Norman Islander BN2B

Islander BN2B

The Islander is currently available with two different powerplants: the Avco Lycoming O-540, 260 HP engine and the IO-540, 300 HP engine. Three wide cabin doors with low sills permit easy and rapid loading for both passengers and cargo. The seats may be quickly removed and stowed, instantly converting the aircraft for freight transport. The Islander can be equipped for photo-survey, fire control, fish spotting, and other roles.

Manufacturer and Model	Seats	Powerplants	Fuel Capacity (std/opt. gal)	Weights Gross/ Empty (lbs)	Cruise Speed (mph) 75% @ alt 65% @ alt	Optimum Range (m w 45 min. rsv) 75% @ alt 65% @ alt	Takeoff/ Landing Distance (ft. over 50′ obst)	Rate of Climb Engine out ROC (fpm)	Service Ceiling	Stall Speed (gear/flaps down, mph)
PILATUS-BRITTEN NORMAN Islander II BN 2B-26	10	2 Lyc. O-540-E4C5 260 hp ea.	137	6600/ 3612	169 @ 7,000′ 154 @ 9,000′	825 @ 7,000′ 946 @ 9,000′	1090/ 960	950/ 192	14,600′	49
PILATUS-BRITTEN NORMAN Islander BN 2A-20 (fg)	10	2 Lyc. IO-540-K1B5 300 hp ea.	135	6560/ 3722	170 @ 7,000′ 167 @ 7,000′	690 @ 7,000′ 759 @ 7,000′	1100/ 960	1130/ 200	18,000′	46
PILATUS-BRITTEN NORMAN Islander BN 2A-21 (fg)	10	2 Lyc. IO-540-K1B5 300 hp ea.	196	6600/ 3762	170 @ 7,000′ 167 @ 7,000′	920 @ 7,000′ 1012 @ 7,000′	1100/ 960	1130/ 200	18,000′	46

Pilatus Britten-Norman Islander BN2T

Islander BN2T

The turbine Islander combines the airframe and undercarriage of the standard Islander with Allison turboprop power. A low cabin floor and large doors provide easy access for passengers and loading of cargo. The all-metal fuselage is fully protected against corrosion, a handy feature because many Islanders spend their lives over or near the water. A dual-wheeled main undercarriage enables the airplane to negotiate soft earth, sand, or snow and ice.

Manufacturer and Model	Seats	Powerplants	Fuel Capacity (std/opt. gal)	Weights Gross/ Empty (lbs)	Cruise Speed (mph) 75% @ alt 65% @ alt	Optimum Range (m w 45 min. rsv) 75% @ alt 65% @ alt	Takeoff/ Landing Distance (ft. over 50′ obst)	Rate of Climb Engine out ROC (fpm)	Service Ceiling	Stall Speed (gear/flaps down, mph)
PILATUS-BRITTEN NORMAN Islander BN2T	10	2 Detroit Diesel Allison 250-B17C gas turbine/ turboprop, 320 5hp units	130/ 250	6600/ 4120	195 @ 10,000′ NA	372 @ 10,000′ 814 @ 10,000′	1120/ 1100	1200 NA	25,000′	NA

NA–Not Available

Pilatus Britten-Norman Trislander

Trislander

The airplane includes three direct-driven Lycoming piston engines, which drive Hartzell constant-speed two-blade fully feathering propellers. High wing propeller clearance reduces blade damage. Automatic propeller feathering is standard and a standby rocket engine (SRE), available as an option, improves performance in the event of an engine failure during takeoff. Up to sixteen passengers can be carried using limousine styled seats.

Manufacturer and Model	Seats	Powerplants	Fuel Capacity (std/opt. gal)	Weights Gross/ Empty (lbs)	Cruise Speed (mph) 75% @ alt 65% @ alt	Optimum Range (m w 45 min. rsv) 75% @ alt 65% @ alt	Takeoff/ Landing Distance (ft. over 50′ obst)	Rate of Climb Engine out ROC (fpm)	Service Ceiling	Stall Speed (gear/flaps down, mph)
PILATUS-BRITTEN NORMAN Trislander BN 2A-Mark III-2 (fg)	18	3 Lyc. O-540-E4C5 260 hp ea.	196	10,000/ 5600	177 @ 7,000′ 172 @ 7,000′	943 @ 7,000′ 1001 @ 7,000′	1919/ 1430	1000/ 145	12,400′	58

PIPER

Piper Turbo Seminole

Seminole and Turbo Seminole

Two hundred pounds of luggage fit easily into this airplane's 24-cubic-foot storage compartment. Optional three-bladed props provide extra ground clearance while taxiing. Fuel drains are in one convenient location for easy preflight inspection.

Manufacturer and Model	Seats	Powerplants	Fuel Capacity (std/opt. gal)	Weights Gross/ Empty (lbs)	Cruise Speed (mph) 75% @ alt 65% @ alt	Optimum Range (m w 45 min. rsv) 75% @ alt 65% @ alt	Takeoff/ Landing Distance (ft. over 50′ obst)	Rate of Climb Engine out ROC (fpm)	Service Ceiling	Stall Speed (gear/flaps down, mph)
PIPER Seminole	4	2 Lyc. O-360-E1A6D 180 hp ea.	110	3800/ 2354	191 @ 8,000′ 186 @ 11,900′	898 @ 7,000′ 979 @ 10,700′	1400/ 1190	1340/ 217	17,100′	68
PIPER Turbo Seminole	4	2 Lyc. O-360-E1A6D & LO-360-E1A6D 180 hp ea.	110	3925/ 2430	211 @ 8,000′ 198 @ 10,700′	904 @ 8,000′ 921 @ 10,700′	1500/ 1190	1290/ 180	20,000′	70

NA–Not Available

Nose and aft storage compartments carry 200 pounds of baggage. Extra-wide cabin door makes for easy entry and exit. Twin turbocharged HP Continental engines give high performance with low maintenance. Power-setting chart and checklists are located on the sun visor for convenience. Optional conference seating has a refreshment cooler that doubles as an extra-wide armrest.

Seneca II

Manufacturer and Model	*Seats*	*Powerplants*	*Fuel Capacity (std/opt. gal)*	*Weights Gross/ Empty (lbs)*	*Cruise Speed (mph) 75% @ alt 65% @ alt*	*Optimum Range (m w 45 min. rsv) 75% @ alt 65% @ alt*	*Takeoff/ Landing Distance (ft. over 50′ obst)*	*Rate of Climb Engine out ROC (fpm)*	*Service Ceiling*	*Stall Speed (gear/flaps down, mph)*
PIPER Seneca II	6	2 Cont. TSIO-360-E 200 hp ea.	98/128	4570/ 2841	218 @ 20,000′ 208 @ 24,000′	628 @ 20,000′ 639 @ 20,000′	1240/ 2090	1340/ 225	25,000′	70

Piper Seneca II

Aerostar (601B and 601P)

The Aerostar's cockpit offers a pilot maximum efficiency and comfort. Center console conveniently groups electric elevator and rudder trim, power steering, autopilot selector panel and, in the 601P, pressurization controls. Five custom color and texture combinations are offered with a choice of soft fabrics or leather. An optional fold-down work table makes business travel easier and more convenient.

Manufacturer and Model	Seats	Powerplants	Fuel Capacity (std/opt. gal)	Weights Gross/ Empty (lbs)	Cruise Speed (mph) 75% @ alt 65% @ alt	Optimum Range (m w 45 min. rsv) 75% @ alt 65% @ alt	Takeoff/ Landing Distance (ft. over 50′ obst)	Rate of Climb Engine out ROC (fpm)	Service Ceiling	Stall Speed (gear/flaps down, mph)
PIPER Aerostar 601B	6	2 Lyc. IO-540-S1A5 290 hp ea.	174	6000/ 3985	295 @ 25,000′ 274 @ 25,000′	NA 1178 @ 15,000′	2490/ 2030	1920/ 470	30,000′	87
PIPER Aerostar 601P	6	2 Lyc. IO-540-S1A5 290 hp ea.	174	6000/ 4056	295 @ 25,000′ 274 @ 25,000′	NA 1178 @ 15,000′	2490/ 2030	2000/ 470	25,000′	87

NA–Not Available

Piper Aerostar 601P

Navajo and Navajo C/R

The maximum speed of the Navajo is 227 kts.; the maximum speed of the Navajo C/R is 228 kts. Both airplanes can transport a significant amount of luggage. The Navajo can carry 200 pounds at one end and another 150 pounds in the nose compartment. With the Navajo C/R, owners can add another 300 pounds of baggage. Individual courtesy lights and oxygen outlets are grouped for easy access. The cockpit and cabin are separated, ensuring privacy.

Piper Navajo

Manufacturer and Model	*Seats*	*Powerplants*	*Fuel Capacity (std/opt. gal)*	*Weights Gross/ Empty (lbs)*	*Cruise Speed (mph) 75% @ alt 65% @ alt*	*Optimum Range (m w 45 min. rsv) 75% @ alt 65% @ alt*	*Takeoff/ Landing Distance (ft. over 50′ obst)*	*Rate of Climb Engine out ROC (fpm)*	*Service Ceiling*	*Stall Speed (gear/flaps down, mph)*
PIPER Navajo	6	2 Lyc. TIO-540-A2C 310 hp ea.	192	6500/ 4003	247 @ 22,000′ 235 @ 24,000′	1157 @ 20,000′ 1214 @ 20,000′	2095/ 1818	1220/ 245	24,000′	80
PIPER Navajo C/R	6	2 Lyc. TIO-540- F2BD 325 hp ea.	192	6500/ 4099	253 @ 20,000′ 239 @ 20,000′	1082 @ 20,000′ 1150 @ 20,000′	2080/ 1818	1200/ 255	24,000′	80

Piper Navajo C/R

Piper Chieftan

Chieftain

The cabin of the Chieftain allows six to ten people to stretch out in comfort on contoured seats upholstered with designer fabrics. The cockpit, separated from the cabin by a curtain, remains private and quiet. Fold-away tables help make business travel easier, and a thermal refreshment center is an optional feature many business and nonbusiness travelers desire.

Manufacturer and Model	Seats	Powerplants	Fuel Capacity (std/opt. gal)	Weights Gross/ Empty (lbs)	Cruise Speed (mph) 75% @ alt 65% @ alt	Optimum Range (m w 45 min. rsv) 75% @ alt 65% @ alt	Takeoff/ Landing Distance (ft. over 50′ obst)	Rate of Climb Engine out ROC (fpm)	Service Ceiling	Stall Speed (gear/flaps down, mph)
PIPER 85	ftain	6	2 Lyc. TIO-540-J2BD 350 hp ea.	192	7000/ 4221	254 @ 20,000′ 242 @ 20,000′	1068 @ 20,000′ 1064 @ 20,000′	2450/ 1880	1120/ 230	24,000′

D-1 Derringer

The Derringer has two Lycoming engines, model IO-320-B1C or CIA, and two Hartzell HC-C2YL/8450-18 propellers. The baggage maximum is 250 pounds. Special features include emergency locator transmitter, stainless steel control cables, and flush tip navigation lights.

Wing D-1 Derringer

Manufacturer and Model	Seats	Powerplants	Fuel Capacity (std/opt. gal)	Weights Gross/ Empty (lbs)	Cruise Speed (mph) 75% @ alt 65% @ alt	Optimum Range (m w 45 min. rsv) 75% @ alt 65% @ alt	Takeoff/ Landing Distance (ft. over 50′ obst)	Rate of Climb Engine out ROC (fpm)	Service Ceiling	Stall Speed (gear/flaps down, mph)
WING D-1 Derringer		2 Lyc. IO-320-B1C, 160 hp.	87	3050/ 2100	NA 210 @ 10,000′	NA 1155 @ 10,000′	1240/ NA	1700/ NA	19,600′	72

NA-Not Available

Turboprops and Turbojets

Atlantic Aviation	P.O. Box 1709 Wilmington, Delaware 19899	Westwind 1 (Jet) Westwind 2 (Jet) Astra (Jet)
Beechcraft	9709 East Central Wichita, Kansas 67201	King Air C90 (Prop) King Air E90 (Prop) King Air B100 (Prop) Super King Air (Prop) King Air F90 (Prop)
Cessna	P.O. Box 1521 Wichita, Kansas 67201	Conquest (Prop) Corsair (Prop) Citation I (Jet) Citation II (Jet) Citation III (Jet)
De Havilland Aircraft	Downsview, Ontario M3K 1Y5 Canada	Dash 7 (Prop) Dash 8 (Jet) Twin Otter (Prop)
Falcon	90 Moonacie Avenue Teterboro, New Jersey 07605	10 (Jet) 200 (Jet) 50 (Jet) 20F (Jet)
Gates Learjeat	Aircraft Division P.O. Box 7707 Wichita, Kansas 67277	25D (Jet) and Modification Longhorn 28 and 29 (Jet) 35A and 36A (Jet) 55 and 56 (Jet)
Government Aircraft Factories	Fishermen's Bend 226 Lorimer Street Port Melbourne, Vic. Australia	Nomad 24A Commuterliner (Prop) Nomad 22B Commuterliner (Prop)

Gulfstream American	P.O. Box 2206 Savannah, Georgia 31402 5001 North Rockwell Avenue Bethany, Oklahoma 73008	Gulfstream III (Jet) 840 (Prop) 980 (Prop) 1000 (Prop) 900 (Prop)
Lear Fan	Box 6000 Reno, Nevada 89506	Lear Fan 2100 (Jet)
Mitsubishi	Suite 1310 12700 Park Central Drive Dallas, Texas 75251	Solitaire (Prop) Marquise (Prop) Diamond I (Jet)
Pilatus	6370 Stans/NW Switzerland	PC-6 Turbo Porter (Prop)
Piper	Lock Haven, Pennsylvania 17745	Cheyenne I (Prop) Cheyenne II (Prop) Cheyenne III (Prop)
Fairchild-Swearingen	P.O. Box 32486 San Antonio, Texas 78284	Merlin IIIC (Prop) Merlin IVC (Prop) Metro III (Prop) Saab-Fairchild 340 (Prop)

ATLANTIC AVIATION

Atlantic Aviation Westwind I

Westwind I

The Westwind 1, capable of 2,800 statute miles, can carry seven passengers nonstop coast to coast. Fuel efficient Garrett turbofans, a sizable fuel capacity, and a generous 1,400-pound payload combine for extra margins of range/payload flexibility. The airplane flies at 41,000 feet, above most weather and most other traffic. An RCA weather radar on board employs a large 18-inch antenna to scan the skies 300 miles ahead. Business travelers will find a refreshment bar offering hot and cold drinks and light snacks, and an optional warming oven for more substantial inflight meals.

Manufacturer and Model	*Seats*	*Powerplants*	*Fuel Capacity (gal)*	*Weights Gross/ Empty (lbs)*	*Optimum Speed (mph)*	*Optimum Range (m w 45 min. rsv @ alt)*	*Takeoff/ Landing Distance (over 50′ obst)*	*Rate of Climb/ Engine out ROC (fpm)*	*Service Ceiling*	*Stall Speed (gear/flaps down, mph)*
ATLANTIC AVIATION Westwind I (turbojets)	10–12	Garrett AiResearch TFE 731-3-1G Turbofan 3700 lbs. ea.	1300	23,000/ 12,800	542 @ 19,400′	2806 @ 41,000′	4950/ 2450	4000/ 1333	45,000′	114

Westwind 2

A pair of powerful, fuel-efficient Garrett turbofan engines provide power for the airplane. Each TFE 731-3 turbofan is sea-level rated at 3,700 pounds thrust for power and efficiency, all the way up to 45,000 feet. Westwind can carry a 1,424-gallon fuel supply, which gives the plane the ability to fly coast to coast or from the West Coast to Hawaii. Inside, travelers have unobstructed views through large windows and can enjoy quiet because engines are located well behind them.

Manufacturer and Model	Seats	Powerplants	Fuel Capacity (gal)	Weights Gross/ Empty (lbs)	Optimum Speed (mph)	Optimum Range (m w 45 min. rsv @ alt)	Takeoff/ Landing Distance (over 50′ obst)	Rate of Climb/ Engine out ROC (fpm)	Service Ceiling	Stall Speed (gear/flaps down, mph)
ATLANTIC AVIATION Westwind 2 (turbojets)	12	2 Garrett TFE-731-3-1G 3700 lbs. ea.	1424	23650/ 13250	543 @ 19,500′	2905 @ 41,000′	5250/ 2450	NA/ NA	45,000′	114

NA–Not Available

Atlantic Aviation Westwind 2

Astra

The Astra, which will be slightly larger than the current Westwind, will be ready by mid-1985. The Astra will have a high-efficiency wing, designed by Israel Aircraft Industries, and will be powered by two Garrett TFE731-100 engines. Cruise speed will be Mach 0.80. Cabin volume will measure 206 inches on internal length, with a height of 67 inches and a width of 57 inches. A 55-cubic-foot baggage compartment is separated from the passenger area.

Specifications and performance data are not yet available.

BEECHCRAFT

King Air C90

Two Pratt & Whitney PT-6A-21 reverse flow, free turbine engines rated at 550 shaft HP each power the plane. An extensive package of avionics equipment, landing gear and brakes, controls, engine instruments, flight instruments, electrical equipment, and lights make the aircraft a popular business and pleasure jet.

Manufacturer and Model	Seats	Powerplants	Fuel Capacity (std/opt. gal)	Weights Gross/ Empty (lbs)	Cruise Speed Max/Econ @ alt (mph)	Optimum Range (m w 45 min. rsv)	Takeoff/ Landing Distance (ft. over 50′ obst)	Rate of Climb/ Engine out ROC (fpm)	Service Ceiling	Stall Speed (gear/flaps down, mph)
BEECH King Air C90 (turboprops)	6-10	2 P&W PT6A-21 550 shp ea.	384	9650/ 5778	255 @ 12,000′ 248 @ 21,000′	1474 @ 21,000′	2261/ 1672	1955/ 539	30,700′	87

Beechcraft King Air C90

Special features of this airplane include auxiliary 27,000 BTU electric heating system (equipped for preheating on the ground), internal corrosion proofing system, low profile glareshield, de-icing equipment, propeller anti-icing, heated windshields, dual pilot and static systems, battery-charging current sensor, safety sentinel, emergency locator transmitter, and good soundproofing system.

King Air E90

Manufacturer and Model	*Seats*	*Powerplants*	*Fuel Capacity (std/opt. gal)*	*Weights Gross/ Empty (lbs)*	*Cruise Speed Max/Econ @ alt (mph)*	*Optimum Range (m w 45 min. rsv)*	*Takeoff/ Landing Distance (ft. over 50′ obst)*	*Rate of Climb/ Engine out ROC (fpm)*	*Service Ceiling*	*Stall Speed (gear/flaps down, mph)*
BEECH King Air C90 (turboprops)	6-10	2 P&W PT6A-21 550 shp ea.	384	9650/ 5778	255 @ 12,000′ 248 @ 21,000′	1474 @ 21,000′	2261/ 1672	1955/ 539	30,700′	87

Beechcraft King Air E90

King Air B100

In addition to a good package of avionics equipment and AiResearch TPE-331-6-252B fixed shaft turbine engines, the King Air B100 features cabin amenities of fresh air outlets for all occupants, ashtrays at every seat, cabin windows with polarized sun shades, "no smoking" and "fasten seat belt" signs with audible chimes, wall-to-wall carpet, cupholder for each seat, sliding doors separating cockpit from cabin, private lavatory, cabin coat cable with hangers, air stair door with hydraulic folding steps.

Beechcraft King Air B100

Manufacturer and Model	Seats	Powerplants	Fuel Capacity (std/opt. gal)	Weights Gross/ Empty (lbs)	Cruise Speed Max/Econ @ alt (mph)	Optimum Range (m w 45 min. rsv)	Takeoff/ Landing Distance (ft. over 50′ obst)	Rate of Climb/ Engine out ROC (fpm)	Service Ceiling	Stall Speed (gear/flaps down, mph)
BEECH King Air B100 (turboprops)	8-15	2 Garrett TPE-331-6-252B 715 shp ea.	470	11,800/ 7112	305 @ 12,000′ 301 @ 21,000′	1525 @ 21,000′	2694/ 2679	2139/ 501	30,430′	96

With one of its Pratt & Whitney PT6A-41 engines, the Super King Air can reach a service ceiling of 19,510 feet with 12,500 pounds; with two engines and the same weight, it can climb to 32,880 feet. For special emergencies, there is an oxygen system on board with 10 automatic deployment masks and one first aid mask.

Super King Air

Beechcraft Super King Air

Manufacturer and Model	Seats	Powerplants	Fuel Capacity (std/opt. gal)	Weights Gross/ Empty (lbs)	Cruise Speed Max/Econ @ alt (mph)	Optimum Range (m w 45 min. rsv)	Takeoff/ Landing Distance (ft. over 50′ obst)	Rate of Climb/ Engine out ROC (fpm)	Service Ceiling	Stall Speed (gear/flaps down, mph)
BEECH Super King Air (turboprops)	8-15	2 P&W PT6A-41 850 shp ea.	544	12,500/ 7543	328 @ 18,000′ 296 @ 31,000′	2184 @ 35,000′	2579/ 2074	2450/ 740	35,000′	86

King Air F90

Every King Air F90 comes equipped with service gear including information kit, two pitot tube covers, two engine logbooks, aircraft logbook, pilot's checklist, power chart, control lock assembly, sump drain wrench, screwdriver, flight bag, coat hangers, warranty ID card from Beechcraft, propeller restraints, and battery manual.

Manufacturer and Model	Seats	Powerplants	Fuel Capacity (std/opt. gal)	Weights Gross/ Empty (lbs)	Cruise Speed Max/Econ @ alt (mph)	Optimum Range (m w 45 min. rsv)	Takeoff/ Landing Distance (ft. over 50′ obst)	Rate of Climb/ Engine out ROC (fpm)	Service Ceiling	Stall Speed (gear/flaps down, mph)
BEECH King Air F90 (turboprops)	6-10	2 P&W PT6A-135 750 shp ea.	470	10,950/ 6640	307 @ 12,000′ 289 @ 26,000′	1813 @ 26,000′	2856/ 2224	2038/ 600	31,000′	87

Beechcraft King Air F90

CESSNA

Conquest

The maximum certified altitude for the Conquest propjet has been increased from 33,000 feet to 35,000 feet, effective with serial number 173. At the new certified maximum altitude, cruise fuel flows are decreased and range is extended. Left-hand seat tracks have been redesigned to allow the left rear seat to slide completely clear of the entry door area, simplifying passenger boarding. The plane can seat 11.

Cessna Conquest

Manufacturer and Model	Seats	Powerplants	Fuel Capacity (std/opt. gal)	Weights Gross/ Empty (lbs)	Cruise Speed Max/Econ @ alt (mph)	Optimum Range (m w 45 min. rsv)	Takeoff/ Landing Distance (ft. over 50′ obst)	Rate of Climb/ Engine out ROC (fpm)	Service Ceiling	Stall Speed (gear/flaps down, mph)
CESSNA Conquest (turboprops)	8-11	2 Garrett TPE 331-8-401S 625 shp ea.	475	9925/ 5687	339 @ 16,000′ 337 @ 24,000′	2382 @ 33,000′	2465/ 1875	2435/ 715	37,000′	87

Cessna Corsair

Corsair

Powered by Pratt & Whitney PT6A-112 engines, the Corsair has a TBO (time between overhauls) of 3,500 hours, which equals approximately six years of normal flying. Baggage capacity is 1,100 pounds, with 500 pounds accommodated in the aft cabin and 600 pounds in the 33-cubic-foot nose compartment. Co-pilot flight instruments are standard equipment.

Manufacturer and Model	Seats	Powerplants	Fuel Capacity (gal)	Weights Gross/ Empty (lbs)	Optimum Speed (mph)	Optimum Range (m w 45 min. rsv @ alt)	Takeoff/ Landing Distance (over 50′ obst)	Rate of Climb/ Engine out ROC (fpm)	Service Ceiling	Stall Speed (gear/flaps down, mph)
CESSNA Corsair (turbojets)	6-8	2 P&W PT6A-112 620 shp ea.	367	8275/ 4980	296 @ 26,500′ 235 @ 26,500′	1453 @ 26,500′	NA/NA	1888/ 424	34,000′	NA

NA–Not Available.

Citation I

The Citation I can operate at altitudes up to 41,000 feet with a range of 1,326 nautical (1,525 statute) miles at an average cruise speed of 352 knots (405 mph). Two Pratt & Whitney JT15D-1A engines power the plane.

Cessna Citation I

Manufacturer and Model	Seats	Powerplants	Fuel Capacity (gal)	Weights Gross/ Empty (lbs)	Optimum Speed (mph)	Optimum Range (m w 45 min. rsv @ alt)	Takeoff/ Landing Distance (over 50′ obst)	Rate of Climb/ Engine out ROC (fpm)	Service Ceiling	Stall Speed (gear/flaps down, mph)
CESSNA Citation I (turbojets)	7-9	2 P&W JT 15D-1A 2200 lbs thrust ea.	564	12,000/ 6557	405 @ 35,000′	1755 @ 41,000′	2930/ 2270	2680/ 800	41,000′	94

Cessna Citation II

Citation II

The Citation II has up to 11 seats and is powered by two Pratt & Whitney JT15D-4 engines. The plane can climb direct to 41,000 feet and cruise at 365 knots (420 mph) over a distance of 1,825 nautical (2,100 statute) miles.

Manufacturer and Model	Seats	Powerplants	Fuel Capacity (gal)	Weights Gross/ Empty (lbs)	Optimum Speed (mph)	Optimum Range (m w 45 min. rsv @ alt)	Takeoff/ Landing Distance (over 50′ obst)	Rate of Climb/ Engine out ROC (fpm)	Service Ceiling	Stall Speed (gear/flaps down, mph)
CESSNA Citation II (turbojets)	9-11	2 P&W JT15D-4 2500 lbs thrust ea.	714	13,500/ 7181	420 @ 25,000′	2265 @ 43,000′	2900/ 2270	3500/ 910	43,000′	93

Citation III

The Citation III can carry 15 passengers and crew and is certified to operate at an altitude of 51,000 feet. Baggage volume is 84 cubic feet. Garrett AiResearch TFE 731-3B-100S engines power the airplane.

Cessna Citation III

Manufacturer and Model	Seats	Powerplants	Fuel Capacity (gal)	Weights Gross/ Empty (lbs)	Optimum Speed (mph)	Optimum Range (m w 45 min. rsv @ alt)	Takeoff/ Landing Distance (over 50′ obst)	Rate of Climb/ Engine out ROC (fpm)	Service Ceiling	Stall Speed (gear/flaps down, mph)
CESSNA Citation III (turbojets)	10-15	2 Garrett TFE-731-3B-100S 3650 lbs. thrust ea.	991	19,700/ 9325	540 @ 33,000′	2878 @ 49,000′	3900/ 2690	5325/ 1625	51,000′	101

DE HAVILLAND

Dash 7

The Dash 7 seats as many as 50 passengers and has room for a buffet, lavatory, and large aft baggage compartment. Two pilots can sit in the cockpit and two flight attendants can take care of passengers. The airplane is powered by four Pratt & Whitney PT6A-50 engines rated at 1,120 shp at takeoff.

Manufacturer and Model	Seats	Powerplants	Fuel Capacity (std/opt. gal)	Weights Gross/ Empty (lbs)	Cruise Speed Max/Econ @ alt (mph)	Optimum Range (m w 45 min. rsv)	Takeoff/ Landing Distance (ft. over 50′ obst)	Rate of Climb/ Engine out ROC (fpm)	Service Ceiling	Stall Speed (gear/flaps down, mph)
DE HAVIL-LAND DHC-7 Dash 7 Series 100 (turboprops)	53	4 P&W PT6A-50 1120 shp ea.	1480	44,000/ 27,650	265 @ 15,000′ 247 @ 15,000′	1614 @ 20,000′	2260/ 1950	1220/ 720	21,000′	76 STOL aircraft

De Havilland Dash 7

De Havilland Dash 8

Dash 8

The Dash 8, which will enter airline service in mid-1984, will be able to seat 32 to 36 passengers. Intended to service routes of up to 500 miles, the plane will be available in a convertible passenger/cargo model and will feature a cruising speed of more than 300 miles per hour. Powerplants for the plane will be two Pratt & Whitney PT7A-2R turbine engines. Specifications and performance data are not yet available.

De Havilland Twin Otter

Twin Otter

The Twin Otter is a 20-seat commuter airplane that offers the flexibility of accommodating a fewer number of persons, 13, and also offering a utility configuration. Two Pratt & Whitney PT6A-27 engines rated at 620 shp for takeoff power the plane.

Manufacturer and Model	Seats	Powerplants	Fuel Capacity (std/opt. gal)	Weights Gross/ Empty (lbs)	Cruise Speed Max/Econ @ alt (mph)	Optimum Range (m w 45 min. rsv)	Takeoff/ Landing Distance (ft. over 50′ obst)	Rate of Climb/ Engine out ROC (fpm)	Service Ceiling	Stall Speed (gear/flaps down, mph)
DE HAVIL-LAND Twin Otter DAC-6 (turboprops)	22	2-P&W PT6A-27 620 shp	378/ 472	12500/ 7415	187 @ 10,000′ 173 @ 10,000′	892 @ 10,000′	2280/ 1500	1600/ 340	25,000′	66

FALCON

Falcon 10

10

The Falcon 10 can accommodate eight passengers plus two crew members. It has two Garrett TFE 731-2 turbofan engines, rear fuselage mounted, 3,230 pounds maximum thrust at sea level, flat rated to 86° F.

Manufacturer and Model	Seats	Powerplants	Fuel Capacity (gal)	Weights Gross/ Empty (lbs)	Optimum Speed (mph)	Optimum Range (m w 45 min. rsv @ alt)	Takeoff/ Landing Distance (over 50′ obst)	Rate of Climb/ Engine out ROC (fpm)	Service Ceiling	Stall Speed (gear/flaps down, mph)
AVIONS MARCEL DASSAULT-BREGUET AVIATION 10 (turbojets)	10	2 Garrett TFE 731-2C 3230 lbs. thrust ea.	882	18740/ 11200	561 @ 27,500′	2246 @ 5,000′	4500/ 2750	4450/ 1050	45,000′	90

200

The 200 includes such standard equipment as strobe lights, wing root landing lights, nose cone ventilation, air conditioning valve position indicator and single point refueling/defueling system. Optional equipment: auxiliary power unit, stereo, pedestal extension, battery temperature indicating system, smoke detection system.

Manufacturer and Model	Seats	Powerplants	Fuel Capacity (gal)	Weights Gross/ Empty (lbs)	Optimum Speed (mph)	Optimum Range (m w 45 min. rsv @ alt)	Takeoff/ Landing Distance (over 50′ obst)	Rate of Climb/ Engine out ROC (fpm)	Service Ceiling	Stall Speed (gear/flaps down, mph)
AVIONS MARCEL DASSAULT-BREGUET AVIATION 200 (turbojets)	11	2 Garrett AiResearch ATF 3-6-1C 5050 lbs. thrust	1586	30650/ 18370	530 @ 29,000′	2822 @ 29,000′	5700/ 2610	NA	NA	NA

NA–Not Available.

50

The 50 accommodates up to 12 passengers, plus a crew of two or three. There are three Garrett TFE 731-3 turbofan engines, rear fuselage mounted, rated at 3,700 pounds thrust at sea level, static, standard day. The center engine is equipped with a thrust reverser.

Manufacturer and Model	Seats	Powerplants	Fuel Capacity (gal)	Weights Gross/ Empty (lbs)	Optimum Speed (mph)	Optimum Range (m w 45 min. rsv @ alt)	Takeoff/ Landing Distance (over 50′ obst)	Rate of Climb/ Engine out ROC (fpm)	Service Ceiling	Stall Speed (gear/flaps down, mph)
AVIONS MARCEL DAUSSAULT-BREGUET AVIATION 50 (turbojets)	12-14	3 Garrett TFE 731-3 3700 lbs. thrust	2,316	38800/ 20640	530 @ 30,000′	4087 @ 30,000′	4900/ 3600	NA	45,000′	87

NA–Not Available.

Falcon 50

Falcon 20F

20F

The 20F can seat nine or 10 passengers and two crew members. G.E. CF700-2D2 engines power the aircraft, which has a thrust in pounds of 4,500. Height of the airplane is 17′5″; wing span is 53′5″.

Manufacturer and Model	Seats	Powerplants	Fuel Capacity (gal)	Weights Gross/ Empty (lbs)	Optimum Speed (mph)	Optimum Range (m w 45 min. rsv @ alt)	Takeoff/ Landing Distance (over 50′ obst)	Rate of Climb/ Engine out ROC (fpm)	Service Ceiling	Stall Speed (gear/flaps down, mph)
AVIONS MARCEL DASSAULT-BREGUET AVIATION 20F (turbojets)	12	2 G.E. CF700-2D2 4500 lbs thrust ea.	1369	28660/ 17060	527 @ 27,000′	2016 @ 5,000′	4600/ 1975	3650/ 900	42,000′	96

GATES LEARJET

Gates Learjet 25D

25D

The 25D is an eight-passenger plane. Inflight amenities include foldout writing table, telephone, hot and cold food and beverage service, and ready access to luggage (up to 500 pounds) enroute. The plane is also widely used as an air taxi and in movement of priority cargo (with optional three-foot-wide door and easily removable seats). It can also be used as an air ambulance, in air-to-ground photo-mapping, weather modification, atmospheric research and geological surveys.

There is also a modification of the 25D available.

Manufacturer and Model	Seats	Powerplants	Fuel Capacity (gal)	Weights Gross/ Empty (lbs)	Optimum Speed (mph)	Optimum Range (m w 45 min. rsv @ alt)	Takeoff/ Landing Distance (over 50' obst)	Rate of Climb/ Engine out ROC (fpm)	Service Ceiling	Stall Speed (gear/flaps down, mph)
GATES LEARJET 25D (turbojets)	10	2 G.E. CJ610-3A 2950 lbs thrust ea.	910	15,500/ 7650	547 @ 25,000'	1766 @ 47,000'	3937/ 2744	6300/ 1725	51,000'	112

LONGHORN 28 and 29

The standard configuration of the Longhorn 28 is eight people and of the Longhorn 29, six people. For takeoff at maximum gross weight, 3,040 feet of runway are required. Maximum takeoff weight for both aircraft is 15,000 pounds; maximum ramp weight for both is 15,500.

Gates Learjet Longhorn 28/29

Manufacturer and Model	Seats	Powerplants	Fuel Capacity (gal)	Weights Gross/ Empty (lbs)	Optimum Speed (mph)	Optimum Range (m w 45 min. rsv @ alt)	Takeoff/ Landing Distance (over 50′ obst)	Rate of Climb/ Engine out ROC (fpm)	Service Ceiling	Stall Speed (gear/flaps down, mph)
GATES LEARJET 28 (turbojets)	10	2 G.E. CJ610-8A 2950 lbs thrust ea.	699	15,500/ 8268	549 @ 25,000′	1415 @ 51,000′	3040/ 2734	NA/ 1900	51,000′	104
GATES LEARJET 29 (turbojets)	8	2 G.E. CJ610-8A 2950 lbs thrust ea.	802	15,500/ 8224	549 @ 25,000′	1701 @ 51,000′	3040/ 2734	NA/ 1900	51,000′	104

NA–Not Available.

Gates Learjet 35A/36A

35A and 36A

The 35A can seat 10 while the 36A can seat eight. Both are powered by Garrett TFE-731-2 engines, and both can climb to a service ceiling of 45,000 feet.

Manufacturer and Model	Seats	Powerplants	Fuel Capacity (gal)	Weights Gross/ Empty (lbs)	Optimum Speed (mph)	Optimum Range (m w 45 min. rsv @ alt)	Takeoff/ Landing Distance (over 50′ obst)	Rate of Climb/ Engine out ROC (fpm)	Service Ceiling	Stall Speed (gear/flaps down, mph)
GATES LEARJET 35A (turbojets)	10	2 Garrett TFE-731-2 3500 lbs thrust ea.	931	17,250/ 9271	542 @ 25,000′	2787 @ 45,000′	4224/ 2884	4900/ 1500	45,000′	110
GATES LEARJET 36A (turbojets)	8	2 Garrett TFE-731-2 3500 lbs thrust ea.	1110	18,250/ 9270	542 @ 25,000′	3,287 @ 45,000′	4784/ 2884	4525/ 1325	45,000′	110

55 and 56

Capable of flying 10 passengers long distances, the 50 series by Gates offers the comforts of a fine executive office: private aft lavatory, hot and cold meal and beverage service, skyphone, plenty of light to read by and sufficient headroom, hiproom and stretch-out room. The aircraft also has aft baggage area with exterior loading door and heated nose and aft baggage compartments.

Gates Learjet 55/56

Manufacturer and Model	Seats	Powerplants	Fuel Capacity (gal)	Weights Gross/ Empty (lbs)	Optimum Speed (mph)	Optimum Range (m w 45 min. rsv @ alt)	Takeoff/ Landing Distance (over 50′ obst)	Rate of Climb/ Engine out ROC (fpm)	Service Ceiling	Stall Speed (gear/flaps down, mph)
GATES LEARJET 55 (turbojets)	13	2 Garrett TFE-731-3 3770 lbs thrust ea.	1015	19,750/ 10,992	556 @ 27,000′	2857 @ 49,000′	4550/ 2950	4675/ NA	51,000′	113
GATES LEARJET 56 (turbojets)	10	2 Garrett TFE-731-3 3770 lbs thrust ea.	1212	20,750/ 11,119	556 @ 27,000′	3467 @ 49,000′	5130/ 2950	NA/NA	51,000′	113

NA–Not Available.

GOVERNMENT AIRCRAFT FACTORIES

Government Aircraft Factories Nomad 24A Commuterliner

Nomad 24A Commuterliner

The 16-passenger Nomad N24A has wide double cargo doors and separate crew entry doors. A retractable undercarriage and twin high-efficiency Allison 250-B17B free turbine engines developing 400 shp each allow cruise speeds of nearly 200 mph.

Manufacturer and Model	Seats	Powerplants	Fuel Capacity (std/opt. gal)	Weights Gross/ Empty (lbs)	Cruise Speed Max/Econ @ alt (mph)	Optimum Range (m w 45 min. rsv)	Takeoff/ Landing Distance (ft. over 50′ obst)	Rate of Climb/ Engine out ROC (fpm)	Service Ceiling	Stall Speed (gear/flaps down, mph)
GOVERNMENT AIRCRAFT FACTORIES Nomad 24A (turboprops)	16	2 Detroit Allison 250-B17B	1,770	9400/ 5197	195 @ 10,000′	1244 @ 10,000′	1706/ 1380	1313/ 240	23,300′	65

Government Aircraft Factories Nomad N22B Commuterliner

Nomad N22B Commuterliner

The Nomad N22B is a normal category twin turboprop light transport aircraft serving 12 passengers, cargo and special operations such as aerial ambulance, geographic and geophysical survey, forestry patrol, and military surveillance and support. A retractable undercarriage and twin high-efficiency Allison 250-B17B free turbine engines developing 400 shp each allow cruise speeds of nearly 200 mph.

Manufacturer and Model	Seats	Powerplants	Fuel Capacity (std/opt. gal)	Weights Gross/ Empty (lbs)	Cruise Speed Max/Econ @ alt (mph)	Optimum Range (m w 45 min. rsv)	Takeoff/ Landing Distance (ft. over 50′ obst)	Rate of Climb/ Engine out ROC (fpm)	Service Ceiling	Stall Speed (gear/flaps down, mph)
GOVERN-MENT AIR-CRAFT FACTORIES Nomad N22B Commuterliner (turboprops)	12	2 Allison 250-B17B 400 shp.	224/297 Imp, gal.	8500/ 4613	190 @ 10,000′	1232 @ 10,000′	1180/1110 STOL 1050/710	1460/ 240	21,000′	62

GULFSTREAM AMERICAN

Gulfstream American Gulfstream III

Gulfstream III

The Gulfstream III, measuring over 83 feet long in its exterior dimensions, has room for two crew members and 19 passengers. Over 157 pounds can be loaded into the baggage compartment, which has both inside and outside door access.

Manufacturer and Model	Seats	Powerplants	Fuel Capacity (gal)	Weights Gross/ Empty (lbs)	Optimum Speed (mph)	Optimum Range (m w 45 min. rsv @ alt)	Takeoff/ Landing Distance (over 50′ obst)	Rate of Climb/ Engine out ROC (fpm)	Service Ceiling	Stall Speed (gear/flaps down, mph)
GULFSTREAM AMERICAN III (turbojets)	19	2 RRSpey MK511-8 (RB163-25) 10,342 lbs thrust ea.	4133	68700/ 32500	512 @ 45,000′	5129 @ 43,000′	5850/ 3400	3800/ 1200	45,000′	139

The 840 has Garrett TPE 331s with fuel capacity up to 420 gallons. Baggage volume equals almost 75 cubic feet. The airplane can cover more than 2,000 statute miles non stop and is FAA-certified for flights into known icing conditions.

840

Manufacturer and Model	Seats	Powerplants	Fuel Capacity (std/opt. gal)	Weights Gross/ Empty (lbs)	Cruise Speed Max/Econ @ alt (mph)	Optimum Range (m w 45 min. rsv)	Takeoff/ Landing Distance (ft. over 50′ obst)	Rate of Climb/ Engine out ROC (fpm)	Service Ceiling	Stall Speed (gear/flaps down, mph)
GULFSTREAM AMERICAN Jetprop Commander 840 (turboprops)	7-10	2 Garrett TPE 331-5-524K 717 shp ea.	430	10,325/ 6120	334 @ 12,000′	2048 @ 31,000′	1833/ 2030	1003	31,000′	86

This model has Garrett TPE 331 engines and can climb to a ceiling of 40,000 feet. The ramp weight of the plane is more than 10,000 pounds and its equipped weight is 6,727 pounds.

980

Manufacturer and Model	Seats	Powerplants	Fuel Capacity (std/opt. gal)	Weights Gross/ Empty (lbs)	Cruise Speed Max/Econ @ alt (mph)	Optimum Range (m w 45 min. rsv)	Takeoff/ Landing Distance (ft. over 50′ obst)	Rate of Climb/ Engine out ROC (fpm)	Service Ceiling	Stall Speed (gear/flaps down, mph)
GULFSTREAM AMERICAN Jetprop Commander 980 (turboprops)	7-10	2 Garrett TPE 331-10-Solk 717 shp ea.	430	10,325/ 6271	362 @ 22,000′	2071 @ 31,000′	1830/ 2150	1010	31,000′	86

The 1000 is certified to fly to 35,000 feet, above most adverse weather conditions. Fuel capacity is 474 usable gallons and statute miles is 2,305. Inside the cabin, there is room for passengers to relax, for a lot of baggage, and for a separate lavatory compartment for privacy.

1000

Manufacturer and Model	Seats	Powerplants	Fuel Capacity (std/opt. gal)	Weights Gross/ Empty (lbs)	Cruise Speed Max/Econ @ alt (mph)	Optimum Range (m w 45 min. rsv)	Takeoff/ Landing Distance (ft. over 50′ obst)	Rate of Climb/ Engine out ROC (fpm)	Service Ceiling	Stall Speed (gear/flaps down, mph)
GULFSTREAM AMERICAN Commander Jetprop 1000 (turboprops)	9–11	Garrett AiResearch TPE-331-10	474	11250/ NA	331 @ 31,000′	2396	NA	929/ NA	35,000′	NA

NA–Not Available.

Gulfstream American 900

900 Gulfstream American's jetprop 900 was only recently certified.

Manufacturer and Model	Seats	Powerplants	Fuel Capacity (std/opt. gal)	Weights Gross/ Empty (lbs)	Cruise Speed Max/Econ @ alt (mph)	Optimum Range (m w 45 min. rsv)	Takeoff/ Landing Distance (ft. over 50′ obst)	Rate of Climb/ Engine out ROC (fpm)	Service Ceiling	Stall Speed (gear/flaps down, mph)
GULFSTREAM AMERICAN Commander Jetprop 900 (turboprops)	NA	NA	NA	10775/ NA	331 @ 12,000′	2246	1937/ 2698	NA/ 924	31,000′	NA

NA–Not Available.

Gulfstream American 900

Lear Fan 2100

Lear Fan 2100

The maximum certificated operating altitude of the Lear Fan 2100 is 41,000 feet, and it is powered by two Pratt & Whitney PT68-35F turboshaft engines. The fuel system consists of two integral wing tanks. Each fuel tank has a capacity of 850 pounds of fuel. Each engine has a separate feed line from the tank with a shutoff valve in the line. Standard navigation lights are mounted on the wing tips. The standard plan seats a crew of two plus six passengers. There is also a refreshment center and toilet with enclosure. The baggage area measures 50 cubic feet.

Manufacturer and Model	Seats	Powerplants	Fuel Capacity (gal)	Weights Gross/ Empty (lbs)	Optimum Speed (mph)	Optimum Range (m w 45 min. rsv @ alt)	Takeoff/ Landing Distance (over 50′ obst)	Rate of Climb/ Engine out ROC (fpm)	Service Ceiling	Stall Speed (gear/flaps down, mph)
LEAR FAN Lear Fan 2100 (turbojets)	8	2P&W PT68-35F	250	7250/ 4000	NA	NA	NA	NA	41,000′	NA

NA–Not Available.

MITSUBISHI

Solitaire

The Solitaire has two 1,000 HP Garrett AiResearch TPE-331-10-501M engines, with a maximum cruising speed of 370 mph. The maximum range with a 45-minute reserve is 1,840 miles and the certified altitude is 31,000 feet. The plane has eight standard places or a maximum of nine places.

Manufacturer and Model	Seats	Powerplants	Fuel Capacity (std/opt. gal)	Weights Gross/ Empty (lbs)	Cruise Speed Max/Econ @ alt (mph)	Optimum Range (m w 45 min. rsv)	Takeoff/ Landing Distance (ft. over 50′ obst)	Rate of Climb/ Engine out ROC (fpm)	Service Ceiling	Stall Speed (gear/flaps down, mph)
MITSUBISHI Solitaire (turboprops)	7-9	2 Garrett TPE-331-10-501M 727 shp ea.	403	10,470/ 7010	369 @ 20,000′ 360 @ 20,000′	1841 @ 31,000′	1800/ 1950	2350/ 475	35,500′	84

Mitsubishi Solitaire

Marquise

The maximum cruising speed of the Marquise is 355 mph; the certified altitude 31,000 feet; and the maximum range with a 45-minute reserve, 1,600 miles. The plane measures 21′6″ on the interior, roomy enough for nine to 11 seats.

Manufacturer and Model	Seats	Powerplants	Fuel Capacity (std/opt. gal)	Weights Gross/ Empty (lbs)	Cruise Speed Max/Econ @ alt (mph)	Optimum Range (m w 45 min. rsv	Takeoff/ Landing Distance (ft. over 50′ obst)	Rate of Climb/ Engine out ROC (fpm)	Service Ceiling	Stall Speed (gear/flaps down, mph)
MITSUBISHI Marquise (turboprops)	9-11	2 Garrett TPE-331-10-501M 778 shp ea.	403	11,575/ 7650	354 @ 16,000′ 339 @ 20,000′	1605 @ 31,000′	2170/ 2200	2200/ 410	33,000′	87

Mitsubishi Marquise

Mitsubishi Diamond I

Diamond I

The Diamond I has Pratt & Whitney JT15D-4 Turbofan engines, a cruising altitude of 41,000 feet, a maximum speed of 500 mph, and a useful load of 5,340 pounds. There is room for two crew members and plenty of baggage.

Manufacturer and Model	Seats	Powerplants	Fuel Capacity (gal)	Weights Gross/ Empty (lbs)	Optimum Speed (mph)	Optimum Range (m w 45 min. rsv @ alt)	Takeoff/ Landing Distance (over 50′ obst)	Rate of Climb/ Engine out ROC (fpm)	Service Ceiling	Stall Speed (gear/flaps down, mph)
MITSUBISHI Diamond I (turbojets)	9-11	2 P&W JT15D-4 2500 lbs. thrust ea.	646	13,890/ 8050	499 @ 30,000′ 472 @ 39,000′	1801 @ 41,000′	4100/ 2700	3000/ 3000	41,000′+	87

PILATUS

The PC-6 is a single-engine, high-wing airplane powered by a Pratt & Whitney PT6A-27 gas turbine engine with seating for up to a maximum of 11 persons. Versatility makes this aircraft especially desirable. It can be quickly converted to perform a wide range of duties, including operating from land, water (with floats), or snow (with skis).

PC-6 Turbo Porter

Manufacturer and Model	Seats	Powerplants	Fuel Capacity (std/opt. gal)	Weights Gross/ Empty (lbs)	Cruise Speed Max/Econ @ alt (mph)	Optimum Range (m w 45 min. rsv)	Takeoff/ Landing Distance (ft. over 50′ obst)	Rate of Climb/ Engine out ROC (fpm)	Service Ceiling	Stall Speed (gear/flaps down, mph)
PILATUS PC-6 Turbo Porter (turboprops)	8–11	P&W PT6A-27	NA	4848/ 2680	149 @ 10,000′	570	3525/ 3270	1270	28,000′	50

NA–Not available

Pilatus PC-6 Turbo Porter

PIPER

Piper Cheyenne I

Cheyenne I

The Cheyenne offers a number of convenient features to make flight a pleasure for passengers. The entry door has illuminated steps. Overhead signs for seat belts and no smoking light chime when passengers' attention is required. An optional digital display unit showing airspeed, outside temperature, altitude, and time allows a passenger to monitor the progress of the flight. The Cheyenne I also offers a private lavatory, eight-track stereo system, and telephone. A full load of baggage can be carried in two oversize, out-of-the-way compartments designed to hold up to 500 pounds.

Manufacturer and Model	Seats	Powerplants	Fuel Capacity (std/opt. gal)	Weights Gross/ Empty (lbs)	Cruise Speed Max/Econ @ alt (mph)	Optimum Range (m w 45 min. rsv)	Takeoff/ Landing Distance (ft. over 50′ obst)	Rate of Climb/ Engine out ROC (fpm)	Service Ceiling	Stall Speed (gear/flaps down, mph)
PIPER Cheyenne I (turboprops)	6-7	2 P&W PT6A-11 500 shp ea.	308	8700/ 4904	287 @ 12,000′ 272 @ 25,000′	1082 @ 25,000′	2444/ 2223	1750/ 413	28,200′	83

Powerful 620 HP twin turboprops get the Cheyenne II off the ground, fully loaded, in just 1,400 feet. The airplane climbs more than 2,700 feet per minute at full gross weight with a useful load of more than 4,000 pounds. In addition to comfortable seating, the interior features convenient folding tables, hot and cold refreshment units, storage cabinets, lavatory, stereo system, and telephone.

Cheyenne II

Piper Cheyenne II

Manufacturer and Model	Seats	Powerplants	Fuel Capacity (std/opt. gal)	Weights Gross/ Empty (lbs)	Cruise Speed Max/Econ @ alt (mph)	Optimum Range (m w 45 min. rsv)	Takeoff/ Landing Distance (ft. over 50' obst)	Rate of Climb/ Engine out ROC (fpm)	Service Ceiling	Stall Speed (gear/flaps down, mph)
PIPER Cheyenne II (turboprops)	6-8	2 P&W PT6A-28 620 shp ea.	390	9000/ 4980	326 @ 11,000' 288 @ 29,000'	1738 @ 29,000'	1980/ 2480	2710/ 660	31,600'	86

Cheyenne III

Seating for eight is standard on the Cheyenne III when a conference arrangement is used. The airplane also gives a new degree of luggage capacity with a total of more than 76 cubic feet in the nose compartment, cabin compartment, and unique wing lockers in the engine nacelles. Folding tables and storage cabinets are all crafted of canyon teak or classic walnut finish. A digital display keeps passengers informed of airspeed, time, outside temperature, and altitude.

Manufacturer and Model	Seats	Powerplants	Fuel Capacity (std/opt. gal)	Weights Gross/ Empty (lbs)	Cruise Speed Max/Econ @ alt (mph)	Optimum Range (m w 45 min. rsv)	Takeoff/ Landing Distance (ft. over 50′ obst)	Rate of Climb/ Engine out ROC (fpm)	Service Ceiling	Stall Speed (gear/flaps down, mph)
PIPER Cheyenne III (turboprops)	NA	2 P&W PT6A-41 720 shp ea.	400	11,000/ 6240	332 @ 20,000′ 311 @ 33,000′	2347 @ 33,000′	2369/ 2820	2400/ 565	32,800′	95

NA/-Not Available

Piper Cheyenne III

Merlin IIIC

The Merlin IIIC is a 9-to-11 place turboprop business aircraft. It features a maximum speed of more than 300 miles per hour, yet consumes 25 percent less fuel than comparable turboprops. The cabin has been redesigned for optimum use of the 18-foot-long interior. Passenger seats are track mounted with adjustable backs, from full vertical through normal settings to total reclining position. Each window has recessed lighting with wood tambour blinds. All overhead lights, reading lamps, and air vents are flush with the ceiling.

Merlin IIIC

Manufacturer and Model	*Seats*	*Powerplants*	*Fuel Capacity (std/opt. gal)*	*Weights Gross/ Empty (lbs)*	*Cruise Speed Max/Econ @ alt (mph)*	*Optimum Range (m w 45 min. rsv)*	*Takeoff/ Landing Distance (ft. over 50′ obst)*	*Rate of Climb/ Engine out ROC (fpm)*	*Service Ceiling*	*Stall Speed (gear/flaps down, mph)*
FAIRCHILD-SWEARINGEN Merlin IIIC (turboprops)	8-11	2 Garrett TPE-331-IOU-503G 900 shp ea.	648	12,500/ 8150	399 @ 15,000′	2860 @ 26,000′	3650/ 3150	3600/ 580	31,000′	102

Merlin IVC

Merlin IVC

The Merlin IVC has a new, longer 57-foot wingspan, employs more powerful 1,000 shp engines with four blade propellers, redesigned cabin interiors, and many safety features. A new refreshment and entertainment center includes a large buffet cabinet that houses beverage and food storage and preparation, a television and stereo. Adjacent to the buffet is a large couch. Behind this are two club seat sections for eight passengers with tables. To the rear of the cabin is a private toilet, vanity, and lavatory. The large baggage compartment is accessible in flight.

Manufacturer and Model	Seats	Powerplants	Fuel Capacity (std/opt. gal)	Weights Gross/ Empty (lbs)	Cruise Speed Max/Econ @ alt (mph)	Optimum Range (m w 45 min. rsv)	Takeoff/ Landing Distance (ft. over 50′ obst)	Rate of Climb/ Engine out ROC (fpm)	Service Ceiling	Stall Speed (gear/flaps down, mph)
FAIRCHILD-SWEARINGEN Merlin IVC (turboprops)	13-16	2-Garrett AiResearch TPE-331-110-601G	648	14200/ 9100	324 @ 15,000′	1755	2850/ 2532	2440/ 627	30,000′	99

Metro III

Metro III, a 19-passenger turboprop airplane, offers greater payload than previous models, longer range and the best climb and cruise performance in its class because of a longer, stronger wing and new more powerful engines. The Metro III can cruise up to 319 miles per hour. Its useful load of 5,463 pounds means that the plane can transport a full capacity load of 19 passengers, a flight crew of two, and luggage for all a distance of 714 statute miles plus a 45-minute reserve. Inside the 33-foot-long cabin, dual, independent air conditioning and heating systems control temperature. Safety features of note include stainless steel nacelles and engine and cabin fire extinguishers.

Manufacturer and Model	*Seats*	*Powerplants*	*Fuel Capacity (std/opt. gal)*	*Weights Gross/ Empty (lbs)*	*Cruise Speed Max/Econ @ alt (mph)*	*Optimum Range (m w 45 min. rsv)*	*Takeoff/ Landing Distance (ft. over 50′ obst)*	*Rate of Climb/ Engine out ROC (fpm)*	*Service Ceiling*	*Stall Speed (gear/flaps down, mph)*
FAIRCHILD-SWEARINGEN Metro III (turboprops)	19	2-AiResearch TPE-331-110 601G	648	14,200/ 8737	319 @ 10,000′	714	3125/ 3125	2440/ 627	30,000′	99

Fairchild Swearingen Metro III

Saab-Fairchild 340

Major features of this aircraft include a pressurized cabin with stand-up head room, cruise speeds of more than 300 mph and the ability to operate from 4,000-foot airfields. The commuter version carries a full load of 34 persons over four 120-mile legs with appropriate reserves and no refueling.

Manufacturer and Model	Seats	Powerplants	Fuel Capacity (std/opt. gal)	Weights Gross/ Empty (lbs)	Cruise Speed Max/Econ @ alt (mph)	Optimum Range (m w 45 min. rsv)	Takeoff/ Landing Distance (ft. over 50′ obst)	Rate of Climb/ Engine out ROC (fpm)	Service Ceiling	Stall Speed (gear/flaps down, mph)
FAIRCHILD-SWEARINGEN Saab-Fairchild 340 (turboprops)	34	2-GE CT7-5	880	24,088/ NA	313 @ 25,000′	1150	3200/ 3600	2100	NA	NA

NA–Not Available.

Saab-Fairchild

Agricultural Aircraft

Ayres	P.O. Box 3090 Albany, Georgia 31706	500 Gallon Turbo Thrush PZL 600 Thrush 400 Gallon Turbo Thrush P&W 600 Thrush 1200 HP Bull Thrush
Cessna	P.O. Box 1521 Wichita, Kansas 67201	Ag Truck Ag Husky
Eagle	Highway 55, Route 1 Boise, Idaho 83702	300 220
Gulfstream American	P.O. Box 2206 Savannah, Georgia 31402	Ag-Cat B
New Zealand Aerospace Industries	Hamilton Airport, R.D. 2 Hamilton, New Zealand U. S. Representative Frontier Aerospace, Inc. 2751 Temple Avenue Long Beach, California 90806	Cresco 600 Fletcher F.U. 24-954
Piper	Lock Haven, Pennsylvania 17745	Pawnee D-235 Brave 300 Brave 375
Weatherly	2344 San Felipe Road Hollister, California 95023	620 620 TP

AYRES

500 Gallon Turbo Thrush

Owners can choose between a PT6A-34AG or PT6A-17AG engine in the 500 Gallon Turbo Thrush agricultural airplane. Standard equipment items include windshield wiper and washer, navigation and instrument lights, two rotating beacons, 200-amp starter generator, universal spray system with external two-inch stainless steel plumbing and streamlined spray booms with outlets for 68 nozzles. Optional equipment: night working lights, landing lights, cockpit fire extinguisher, air conditioning, instrument installation, avionics installation, high density spray configuration, and five-position adjustable spray nozzles.

Manufacturer and Model	Seats	Powerplants	Hopper Capacity (gal)	Fuel Capacity (std/opt. gal)	Weights Gross/ Empty (lbs)	Working Speed (mph)	Takeoff/ Landing Distance (ft. over 50′ obst)	Rate of Climb (fpm)	Stall Speed (gear/flaps down, mph)
AYRES 500 Gallon Turbo Thrush	NA	P&W PT6A-34AG	500	100	6,000/3,900	95–150	600/ 500	1740	57

NA–Not Available

PZL 600 Thrush

With its new two-place configuration, this model is available with the PZL-3S engine providing 600 HP. Optional equipment includes high density spray configuration, five-position adjustable spray nozzles, 65208 Agrinautics pump.

Manufacturer and Model	Seats	Powerplants	Hopper Capacity (gal)	Fuel Capacity (std/opt. gal)	Weights Gross/ Empty (lbs)	Working Speed (mph)	Takeoff/ Landing Distance (ft. over 50′ obst)	Rate of Climb (fpm)	Stall Speed (gear/flaps down, mph)
AYRES PZL Thrush	2	PZL-R3S 600 hp.	400	106	6900/3700	105–115	775/ 500	900	64

400 Gallon Turbo Thrush

There are three engine options with the 400 model: a PT6A-34AG 750 HP, a PT6A-15AG HP, or a PT6A-11AG 500 HP. Standard equipment includes disc brakes, wire cutters on main gear, 24-volt electrical system, and sealed cockpit enclosure.

Manufacturer and Model	Seats	Powerplants	Hopper Capacity (gal)	Fuel Capacity (std/opt. gal)	Weights Gross/ Empty (lbs)	Working Speed (mph)	Takeoff/ Landing Distance (ft. over 50′ obst)	Rate of Climb (fpm)	Stall Speed (gear/flaps down, mph)
AYRES 400 Gallon Turbo Thrush	1	P&W PT6A-15AG 680 hp.	400	106	8200/3600	95–150	815/ 500	1740	57

P&W 600 Thrush

Powered by the 600 HP P&W R-1340 engine, this airplane has a fuel capacity of 100 gallons, a gross weight of 6,000 pounds, and a length of more than 29 feet. Some of the standard equipment items are a universal spray system with external stainless steel plumbing, root model 67 pump, streamlined spray booms with outlets for 68 nozzles, and a side-loading system on the left side.

Manufacturer and Model	Seats	Powerplants	Hopper Capacity (gal)	Fuel Capacity (std/opt. gal)	Weights Gross/ Empty (lbs)	Working Speed (mph)	Takeoff/ Landing Distance (ft. over 50′ obst)	Rate of Climb (fpm)	Stall Speed (gear/flaps down, mph)
AYRES P&W 600 Thrush	1	P&W R-1340 600 hp.	400	106	6900/3700	105–115	775/ 500	900	64

Ayres 1200 HP Bull Thrush

1200 HP Bull Thrush

The Bull Thrush cruises at 159 mph with spray equipment on board. Standard equipment includes wingtip strobe lights, adjustable rudder pedals, windshield wiper and washer, hopper window into the 510 gallon hopper, streamlined spray booms. Optional items are night working lights, landing lights on the left or right wing, cockpit fire extinguisher, five-position adjustable spray nozzles, and instrument and avionics installation.

Manufacturer and Model	Seats	Powerplants	Hopper Capacity (gal)	Fuel Capacity (std/opt. gal)	Weights Gross/ Empty (lbs)	Working Speed (mph)	Takeoff/ Landing Distance (ft. over 50′ obst)	Rate of Climb (fpm)	Stall Speed (gear/flaps down, mph)
AYRES 1200 HP Bull Thrush	NA	Wright R-1820 1200 hp.	510	190	6000/4990	100–150	550/ 550	2,033	57

NA–Not available

CESSNA

Cessna Ag Truck

Ag Truck

Ag Truck models have a 280-gallon hopper and a 300 HP engine for an efficient combination when pilot applies either liquid or dry materials, at heavy or low rates on large and small fields. The airplane's hydraulic dispersal system assures constant boom pressure at any speed without the drag of a fan-powered system. Available options for the Ag Truck include air conditioning and a six-way articulating seat.

Manufacturer and Model	Seats	Powerplants	Hopper Capacity (gal)	Fuel Capacity (std/opt. gal)	Weights Gross/ Empty (lbs)	Working Speed (mph)	Takeoff/ Landing Distance (ft. over 50′ obst)	Rate of Climb (fpm)	Stall Speed (gear/flaps down, mph)
CESSNA Ag Truck	1	Cont. IO-520-D 300 hp.	280	54	4200/2222	121	2140/ 1265	480	65

The turbocharged Ag Husky has a 310 HP engine providing full-time power at application altitudes. Standard equipment includes a 280-gallon hopper, hydraulic dispersal system, special lighting package, full flow air filter, three-bladed propeller. All Cessna agricultural planes have new exterior styling with new color and stripe combinations. Interior styling features two-color combination seat styling with all vinyl fabric and optional Reevene upholstery.

Ag Husky

Cessna Ag Husky

Manufacturer and Model	***Seats***	***Powerplants***	***Hopper Capacity (gal)***	***Fuel Capacity (std/opt. gal)***	***Weights Gross/ Empty (lbs)***	***Working Speed (mph)***	***Takeoff/ Landing Distance (ft. over 50′ obst)***	***Rate of Climb (fpm)***	***Stall Speed (gear/flaps down, mph)***
CESSNA Ag Husky	1	Cont. TSIO-520-T 310 hp.	280	54	4400/2293	121	1975/ 1265	510	61

EAGLE

300

The 55-foot wingspan of the 300 offers short turns and wide swath so that the pilot can make fewer passes and the airplane can consume less fuel. The canopy provides excellent visibility in all quadrants and seals the cockpit from toxic chemicals. Ram air charcoal-filtered ventilation system provides clean, fresh air for the pilot and access windows for inflight windshield cleaning.

Manufacturer and Model	Seats	Powerplants	Hopper Capacity (gal)	Fuel Capacity (std/opt. gal)	Weights Gross/ Empty (lbs)	Working Speed (mph)	Takeoff-Landing Distance (ft. over 50′ obst)	Rate of Climb (fpm)	Stall Speed (gear/flaps down, mph)
EAGLE									
300 Biplane	1	Lyc. IO-540-M1A5D	250	41 (28 auxiliary)	5400/2322	65–120	NA/NA	NA	51 (power on)

NA–Not Available

Eagle 300

Eagle 220

The 220 has a Continental engine, Hartzell prop, and wingspan of 55 feet. The airplane is equipped to distribute chemicals efficiently and uniformly with a minimum of loss through drift or evaporation. Fuel capacity is 40 gallons.

220

Manufacturer and Model	*Seats*	*Powerplants*	*Hopper Capacity (gal)*	*Fuel Capacity (std/opt. gal)*	*Weights Gross/ Empty (lbs)*	*Working Speed (mph)*	*Takeoff/ Landing Distance (ft. over 50′ obst)*	*Rate of Climb (fpm)*	*Stall Speed (gear/flaps down, mph)*
EAGLE 220 Biplane	1	Continental W-670-6N	250	41 (28 auxiliary)	5100/2549	65–95	NA/NA	NA	51 (power on)

NA–Not Available

GULFSTREAM AMERICAN

Ag-Cat B

Since corrosion is a major problem in ag-planes, Gulfstream American designed the Ag-Cat B with easily removable aluminum alloy panels so it can be washed out quickly and thoroughly. A choice of dispersal systems match the widest range of aerial application requirements. The booms on the plane have provisions for 69 evenly spaced nozzles. To convert to solids dispersal, the entire spray system can be quickly removed without special tools. Aerodynamic spreaders for dry materials produce wide, even swath patterns at application rates that can vary from just a few pounds per acre to as much as several hundred pounds per acre. If obstructions occur, the entire hopper load, either liquids or solids, can be dumped in a few seconds.

Gulfstream American AG-CAT B

Manufacturer and Model	*Seats*	*Powerplants*	*Hopper Capacity (gal)*	*Fuel Capacity (std/opt. gal)*	*Weights Gross/ Empty (lbs)*	*Working Speed (mph)*	*Takeoff/ Landing Distance (ft. over 50′ obst)*	*Rate of Climb (fpm)*	*Stall Speed (gear/flaps down, mph)*
GULFSTREAM AMERICAN Ag-Cat B/450	1	P&W R-985 450 hp.	300	46/80	6075/3100	105–109	1090/ 933	610	53
GULFSTREAM AMERICAN Ag-Cat B/600	1	P&W R-1340 600 hp.	300	46/80	6075/3255	109–120	850/ 1054	797	54

NEW ZEALAND AEROSPACE INDUSTRIES

Cresco 600

Cresco 600

Powering the Cresco is the Avco Lycoming LTP 101, a fuel-efficient turboprop. Standard equipment includes adjustable rudder pedals, electric elevator trim, front passenger seat, large rear cargo door, tinted windshield, and canopy side windows. Optional equipment: dual flight controls, cockpit heating system, and a range of dispersal systems from ultra-high volume solids through to an ultra-low volume spray system.

Manufacturer and Model	Seats	Powerplants	Hopper Capacity (gal)	Fuel Capacity (std/opt. gal)	Weights Gross/ Empty (lbs)	Working Speed (mph)	Takeoff/ Landing Distance (ft. over 50′ obst)	Rate of Climb (fpm)	Stall Speed (gear/flaps down, mph)
NEW ZEALAND AEROSPACE INDUSTRIES Cresco 600	1	AVCO Lycoming LTP 101/600A-1A	470	NA	7000 (agricultural)/ 2560	146	NA/NA	745	52

NA–Not Available.

Fletcher F.U. 24–954

Apart from its agricultural role, the aircraft can be configured for such tasks as air ambulance, passenger carrier for six or eight, cargo hauler with 119-cubic-foot capacity, aerial survey and photography, and many other roles. An easily operated "dump" mechanism enables the pilot to release his complete load within five seconds in case of emergency, and the nosewheel absorbs a large portion of the impact forces in an accident.

Fletcher F.U. 24-954

Manufacturer and Model	Seats	Powerplants	Hopper Capacity (gal)	Fuel Capacity (std/opt. gal)	Weights Gross/ Empty (lbs)	Working Speed (mph)	Takeoff/ Landing Distance (ft. over 50′ obst)	Rate of Climb (fpm)	Stall Speed (gear/flaps down, mph)
NEW ZEALAND AEROSPACE INDUSTRIES Fletcher F.U. 24–954	1–8	Lycoming 10–720 400 hp.	2320 lbs.	NA	5430–2620	130	1220/ 1280	920	56

NA–Not Available.

Piper Pawnee D

Pawnee D-235

The Piper Pawnee D, like other Piper ag-planes, can be engaged in controlling seeding, fertilizing, and applying solid and liquid chemicals. Advantages over ground equipment include faster crop coverage, with effective dispersal of solids and liquids at 6–7 acres per minute; ability to work when ground equipment cannot due to maturity and size of crops or wet soil conditions; safety of the carefully controlled and filtered cockpit environment. The Pawnee D is powered by a 235 HP low-compression Lycoming engine with six cylinders and dual ignition. The fiberglass multipurpose hopper has a capacity of 150 gallons of liquids or 1,200 pounds of solids.

Manufacturer and Model	*Seats*	*Powerplants*	*Hopper Capacity (gal)*	*Fuel Capacity (std/opt. gal)*	*Weights Gross/ Empty (lbs)*	*Working Speed (mph)*	*Takeoff/ Landing Distance (ft. over 50′ obst)*	*Rate of Climb (fpm)*	*Stall Speed (gear/flaps down, mph)*
PIPER Pawnee D	1	Lyc. O-540 235 hp.	150	38	2900/1599	65/115	1350/ NA	700	70

NA–Not Available.

Brave 300

The powerplant of the Brave 300 is the Lycoming six-cylinder, dual ignition engine with Bendix fuel injection and constant speed propeller. Owners have a choice of 30- and 38-cubic-foot hoppers (225 and 275 gallons), and of quick-change spray systems in high, medium, and low volume, plus an all-purpose spreader for solids in a choice of stainless steel or aluminum. An optional high-volume heater in sealed cockpit area eliminates the need for cumbersome clothing on chilly mornings. Plastic loading door is unpainted and hopper is translucent so the pilot can see contents. A clear glass window in back of hopper aids pilot in monitoring contents.

Piper Brave 300

Manufacturer and Model	Seats	Powerplants	Hopper Capacity (gal)	Fuel Capacity (std/opt. gal)	Weights Gross/ Empty (lbs)	Working Speed (mph)	Takeoff/ Landing Distance (ft. over 50′ obst)	Rate of Climb (fpm)	Stall Speed (gear/flaps down, mph)
PIPER Brave 300	1	Lyc. IO-540-K1G5 300 hp.	225/275	86	3900/2198	75/127	1525/ 1650	770	71

Brave 375

An agricultural airplane, the Brave 375 has a Lycoming IO-720-D1CD engine. Hopper capacity is 275 standard gallons and the maximum hopper load is 1,900 standard pounds. Special agricultural features include wire cutter, 38-cubic-foot hopper, corrosion-resistant white paint, corrosion resistant primer, remote reading hopper gauge, sub tank and gate control.

Piper Brave 375

Manufacturer and Model	Seats	Powerplants	Hopper Capacity (gal)	Fuel Capacity (std/opt. gal)	Weights Gross/ Empty (lbs)	Working Speed (mph)	Takeoff/ Landing Distance (ft. over 50′ obst)	Rate of Climb (fpm)	Stall Speed (gear/flaps down, mph)
PIPER Brave 375	1	Lyc. IO-720-DICD 375 hp.	275	86	3900/2465	78/132	1208/ 1850	1051	76

WEATHERLY

620

The tip vanes of the 620 were developed to improve spraying efficiency through two primary functions: to increase the effective swath width and to reduce the amount of spray materials lost from the swath area. The vanes may be folded for hangar storage. They will also fold back if they strike an obstruction so that damage is minimized. The spray boom was designed for minimum drag and easy conversion for solids dispersal.

Weatherly 620

Manufacturer and Model	Seats	Powerplants	Hopper Capacity (gal)	Fuel Capacity (std/opt. gal)	Weights Gross/ Empty (lbs)	Working Speed (mph)	Takeoff/ Landing Distance (ft. over 50′ obst)	Rate of Climb (fpm)	Stall Speed (gear/flaps down, mph)
WEATHERLY 620	1	P&W R-985 450 hp.	335	65	5000/2850	92–120	NA/980	980	66

NA–Not Available.

620 TP

The 620 TP is the 620 airplane with a slightly enlarged hopper and fuel capacity and the powerplant of a Pratt & Whitney engine. Ferrying at 120 mph with a full load and full fuel only requires about 305 HP on a standard day. Maintenance for the aircraft and engine is relatively simple. Easily removed panels provide quick access to all parts of the aircraft. Standard equipment includes 24V electric system, spray system, thrust reversing, and compressor wash. Optional equipment: solids dispersal equipment, wing tip vanes.

Weatherly 620 T P

Manufacturer and Model	Seats	Powerplants	Hopper Capacity (gal)	Fuel Capacity (std/opt. gal)	Weights Gross/ Empty (lbs)	Working Speed (mph)	Takeoff/ Landing Distance (ft. over 50′ obst)	Rate of Climb (fpm)	Stall Speed (gear/flaps down, mph)
WEATHERLY 620TP	1	P&W PT6A-11AG 500 hp.	340	65	5000/2500	92–144	NA/980	980	66

NA–Not Available.

Schleicher Sailplanes	P.O. Box 118 Port Matilda, Pennsylvania 16870	ASW-20
Schweizer	P.O. Box 147 Elmira, New York 14802	1-36 2-33A

SCHLEICHER

ASW-20

The ASW-20 sailplane from Schleicher includes standard equipment of adjustable rudder pedals, adjustable back rest, retractable landing gear, full-vision plexiglass canopy with sliding vent window, fittings for oxygen, tail dolly. Options are tinted blue canopy, headrest, nose statics, tail wheel, trailers, water ballast.

Manufacturer and Model	Seats	Weights Gross/ Empty (lbs)	Max Speed (mph)	Stall Speed (mph)	Airplane Tow (max)	Auto/Winch Tow (max)	Min Sink Speed (mph)	Best Lift Over Drag Speed (mph)	Lowest Rate of Sink (fps)	Glide Ratio
SCHLEICHER ASW-20	1	1000/ 540	161	37	109	75	46	69	2	43:1

1-36

Compact and lightweight, this 46-foot-span sailplane is made of an aluminum structure with standard effective dive brakes, T-tail, large cockpit, and good visibility. Known also as the Sprite, it is offered with two landing gear positions: the 36903-1 version with forward wheel position, and the 36903-3 version with an aft wheel position. The aft wheel position version is recommended for school and club operation where ruggedness and ease of ground handling are important considerations. A fixed main wheel with hydraulic brake is fitted on all planes. The standard color of the plane is white enamel with red trim. Available at no extra cost is a yellow base color with a choice of blue, orange or black trim.

Schweizer 1-36

Manufacturer and Model	Seats	Weights Gross/ Empty (lbs)	Max Speed (mph)	Stall Speed (mph)	Airplane Tow (max)	Auto/Winch Tow (max)	Min Sink Speed (mph)	Best Lift Over Drag Speed (mph)	Lowest Rate of Sink (fps)	Glide Ratio
SCHWEIZER 1-36 (Sprite)	1	710/ 450	123	36	123	78	46	57	2.25	31:1

Schweizer 2-33A

2-33A

The 2-33 was developed from the 2-22, the first two-place sailplane specifically designed for clubs and commercial schools. The ruggedness and strength of the 2-33 begins with the welded chrome-alloy steel tube fuselage and horizontal surfaces. These components are corrosion-proofed with zinc chromate, and then are covered with Ceconite fabric finished with aircraft dope and enamel. The wings and vertical surfaces are of all aluminum construction including the surface skins. The aluminum frame rudder is Ceconite covered. Inside the cockpit, there are molded arm rests in the front seat. The 2-33 is equipped with dual controls. Other special features are the one-piece, free-blown canopy providing excellent visibility for occupants in both seats, and the aerodynamically balanced dive brakes providing good descent and landing control.

Manufacturer and Model	*Seats*	*Weights Gross/ Empty (lbs)*	*Max Speed (mph)*	*Stall Speed (mph)*	*Airplane Tow (max)*	*Auto/Winch Tow (max)*	*Min Sink Speed (mph)*	*Best Lift Over Drag Speed (mph)*	*Lowest Rate of Sink (fps)*	*Glide Ratio*
SCHWEIZER 2-33A	2	1040/ 600	99	32 solo 34 dual	99	69	38 solo 42 dual	46 solo 53 dual	2.6 solo 3.1 dual	23:1

Aerospatiale	1701 West Marshall Drive Grand Prairie, Texas 75051	AS Astar 350C SA Lama 315B SA Gazelle 341 SA Alouette 319B SA Dauphin 360C SA 365N Dauphin 2 AS 332C Super Puma AS 355 TwinStar
Agusta	One West Loop South Houston, Texas 77027	109A Mark II
Bell	P.O. Box 482 Ft. Worth, Texas 76101	206B JetRanger III 206L-I LongRanger II 205A 222 212 214ST 412 214B
Boeing Vertol	P.O. Box 16858 Philadelphia, Pennsylvania 19142	Chinook 234
Brantly-Hynes	P.O. Box 1046 Frederick, Oklahoma 73542	B-2 Series
Enstrom	P.O. Box 277 Menominee, Michigan 49858	F-28C-2 280C Shark F-28F 280F Shark
Hiller	2075 West Scranton Avenue Porterville, California 93257	UH 12E UH 12E4 UH 12ET UH 12E4T UH FH 1100
Hughes	Culver City, California 90230	300C 500D

MBB	P.O. Box 514 West Chester, Pennsylvania 19380	BO 105 CBS BK 117
Robinson	Torrance Municipal Airport Torrance, California 90505	R-22
Rotorway	14805 South Interstate 10 Tempe, Arizona 85281	Exec Scorpion 133
Sikorsky	Division of United Technologies Stratford, Connecticut 06602	S-76 ABC

The Astar 350C combines the economies of fiberglass rotor head and blades, elastomeric bearings, polycarbonate hull, and modular engine, which enables the operator to remove and replace major engine sections in the field. The Astar is also a cargo lifter with a payload of more than a ton. The cabin converts in a few minutes from passenger configuration to unobstructed cargo floor.

AS Astar 350C

Manufacturer and Model	Seats	Powerplant	Fuel (gal, no rsv)	Weights Gross/ Empty (lbs)	Cruise Speed (mph)	Never Exceed Speed (Vne)	Max Range m @ alt	Hover OGE	Hover IGE
AERO-SPATIALE Astar AS 350C	6	Lyc. LTS-101-600A2 615 shp ea.	140	4300/ 2360	147 @ 1,640′	169	491 @ 1,640′	5400′	8,800′

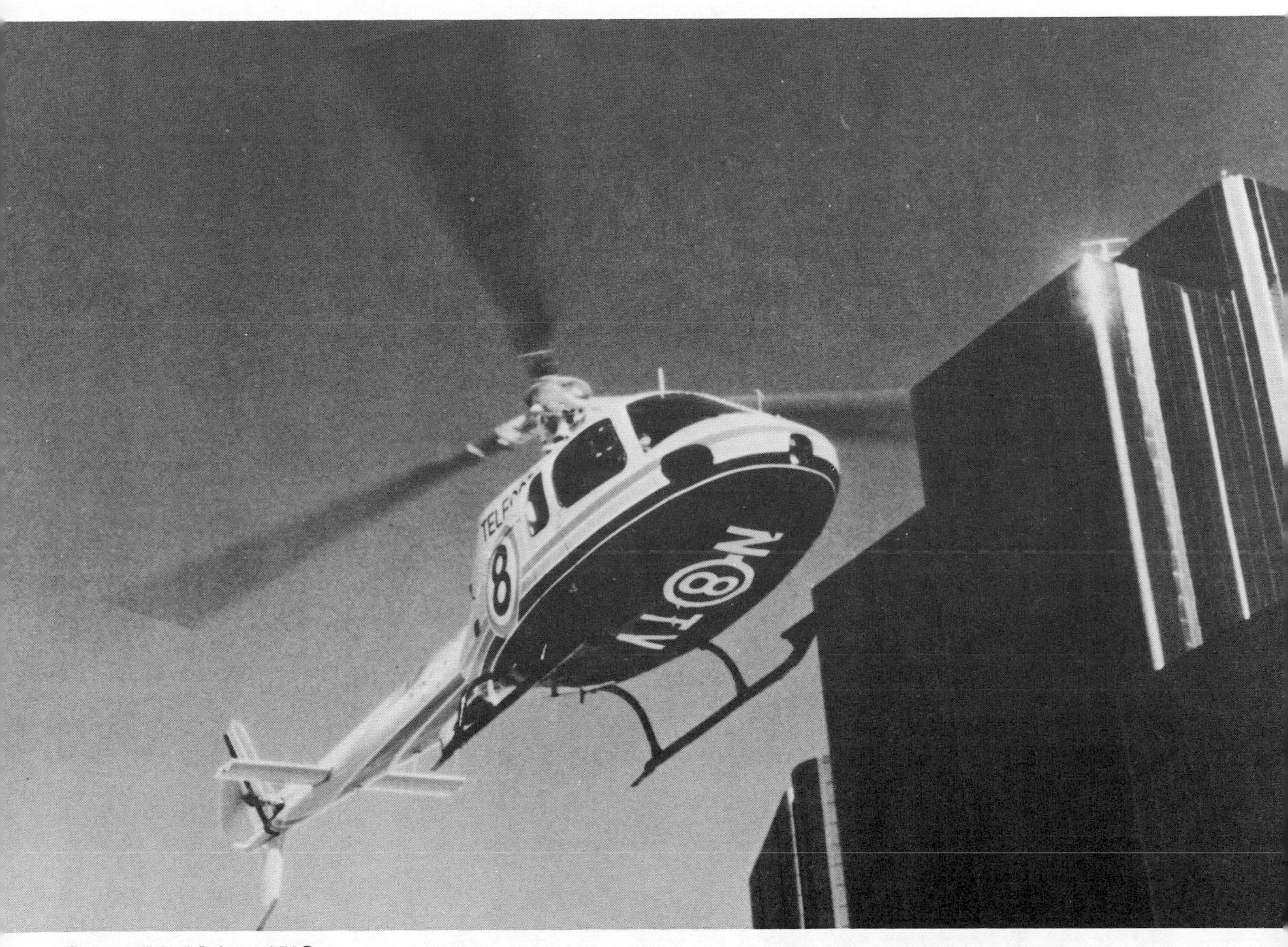

Aerospatiale AS Astar 350C

SA Lama 315B

The Lama can take off from altitudes of 16,000 feet and operate up to 20,000 feet with three men on board and enough fuel for three and a half hours. By just one setting on its unique computer, the flight calculations for density, altitude, power, maximum hover gross weight, and best cruising altitude can be determined prior to lift-off to ensure greater safety during flight. The copter also comes equipped with an aerial crane capable of lifting 2,500 pounds. It can perform agricultural duties.

Manufacturer and Model	Seats	Powerplant	Fuel (gal, no rsv)	Weights Gross/ Empty (lbs)	Cruise Speed (mph)	Never Exceed Speed (Vne)	Max Range m @ alt	Hover OGE	Hover IGE
AERO-SPATIALE Lama SA 315B	5	Turbomeca Artouste IIIB 562 shp	152	4300/ 2215	119 @ 1,640′	130	320 @ 1,640′	15,100′	16,600′

Aerospatiale SA 315B Lama

Aerospatiale SA Gazelle 341

SA Gazelle 341

Gazelle was the first helicopter to be FAA-approved for single-pilot IFR. Now it is certified for CAT II operation. The IFR instrumentation includes a Sperry flight control system that allows a pilot to operate safely in high-density ATC environments. Equipment includes fiberglass, three-bladed rotor, fixed shaft turbine engine. The Gazelle can perform many tasks: offshore work, air ambulance missions, and executive transport.

Manufacturer and Model	Seats	Powerplant	Fuel (gal, no rsv)	Weights Gross/ Empty (lbs)	Cruise Speed (mph)	Never Exceed Speed (Vne)	Max Range m @ alt	Hover OGE	Hover IGE
AERO-SPATIALE SA Gazelle 341	1–4	NA	NA	3970/ 2127	167 @ sea level	224 @ sea level	398 @ sea level	7,215′	9,185′

NA–Not Available

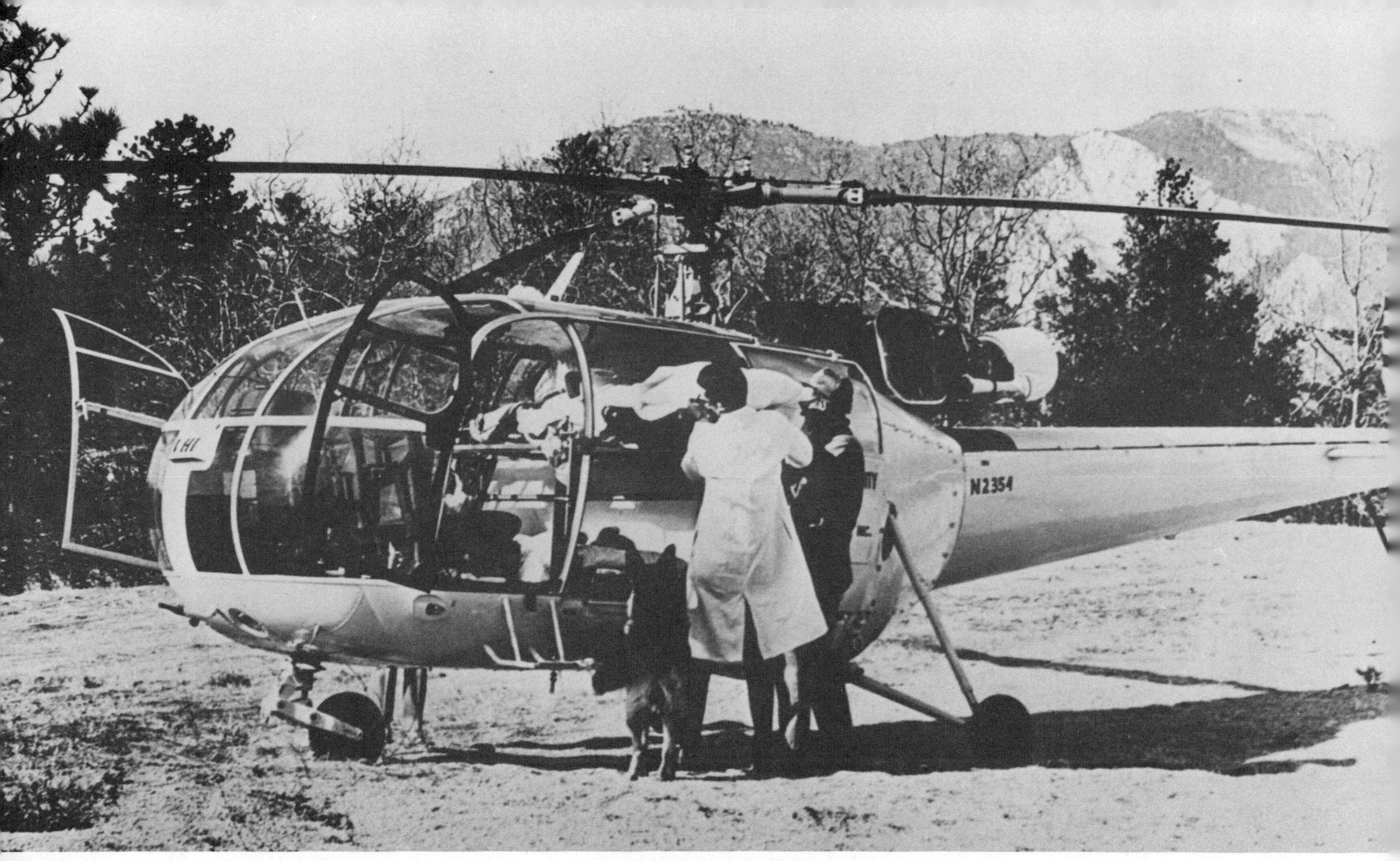

Aerospatiale SA Alouette 319B

SA Alouette 319B

The Alouette is popular for transporting cargo worldwide, passengers, slingloads, rescue, as an ambulance, and for fire fighting work. With its ability to fly into remote areas or hard-to-reach accident sights, the Alouette 3 and its crew can function quickly when ground transportation fails.

Manufacturer and Model	Seats	Powerplant	Fuel (gal, no rsv)	Weights Gross/ Empty (lbs)	Cruise Speed (mph)	Never Exceed Speed (Vne)	Max Range m @ alt	Hover OGE	Hover IGE
AERO-SPATIALE SA Alouette 319B	7	Turbomeca Astazou 14B 562 shp	152	4960/ 2486	126 @ 1,640′	137	350 @ 1,640′	5,574′	10,170′

SA Dauphin 360C

As the successor to the Alouette 319B, the Dauphin features a cruising speed of 168 miles per hour and a range of more than 400 miles. It can accommodate up to 13 passengers plus a pilot in a 177-cubic-foot cabin. There is an additional 35-cubic-foot area for baggage. Its cargo hook can lift 2,866 pounds. The turbine engine offers low fuel consumption, electronic start-up and a high-power ignition system linked to an automatic-sequence control. One of its many options include skis for landing on snow-covered surfaces that will not restrict the use of the standard landing gear.

Manufacturer and Model	Seats	Powerplant	Fuel (gal, no rsv)	Weights Gross/ Empty (lbs)	Cruise Speed (mph)	Never Exceed Speed (Vne)	Max Range m @ alt	Hover OGE	Hover IGE
AERO-SPATIALE SA Dauphin 360C	10-14	Turbomeca Astazou XVIII A 1032 shp	169	6615/ 3797	168 @ 1,640′	195	407 @ 1,640′	5,740′	8,035′

Aerospatiale SA 360C Dauphin

Aerospatiale SA 365N Dauphin 2

SA 365N Dauphin 2

The Dauphin 2 combines all the features of the original Dauphin with a twin-engine package of 725 HP turboshafts to give added power. Accessible through four wide doors, the large-volume cabin offers various layouts: a deluxe version for nine passengers, a VIP version for four, five or six passengers and a utility version for transporting 13 passengers and one pilot over short hauls. With seats removed, the cabin has a flat floor for freight transport or the installation of ambulance equipment. The model can also carry out any missions in inclement weather with its complete IFR set of instruments designed for one or two pilots.

Manufacturer and Model	Seats	Powerplant	Fuel (gal, no rsv)	Weights Gross/ Empty (lbs)	Cruise Speed (mph)	Never Exceed Speed (Vne)	Max Range m @ alt	Hover OGE	Hover IGE
AERO-SPATIALE SA 365N Dauphin 2	14	2 Turbo-meca Arriel 1C 725 hp ea.	291	7935/ 4188	153 @ sea level	195	525 @ sea level	3,445′	6,070′

AS 332C Super Puma

The Super Puma is fitted with two 1,755 turbine engines providing a large power reserve and hot-day, high-altitude, single-engine performance. The cabin can accommodate 20 passengers plus three crew members. Also noteworthy are the IFR instrumentation and weather radar, which combine to provide all-weather capability.

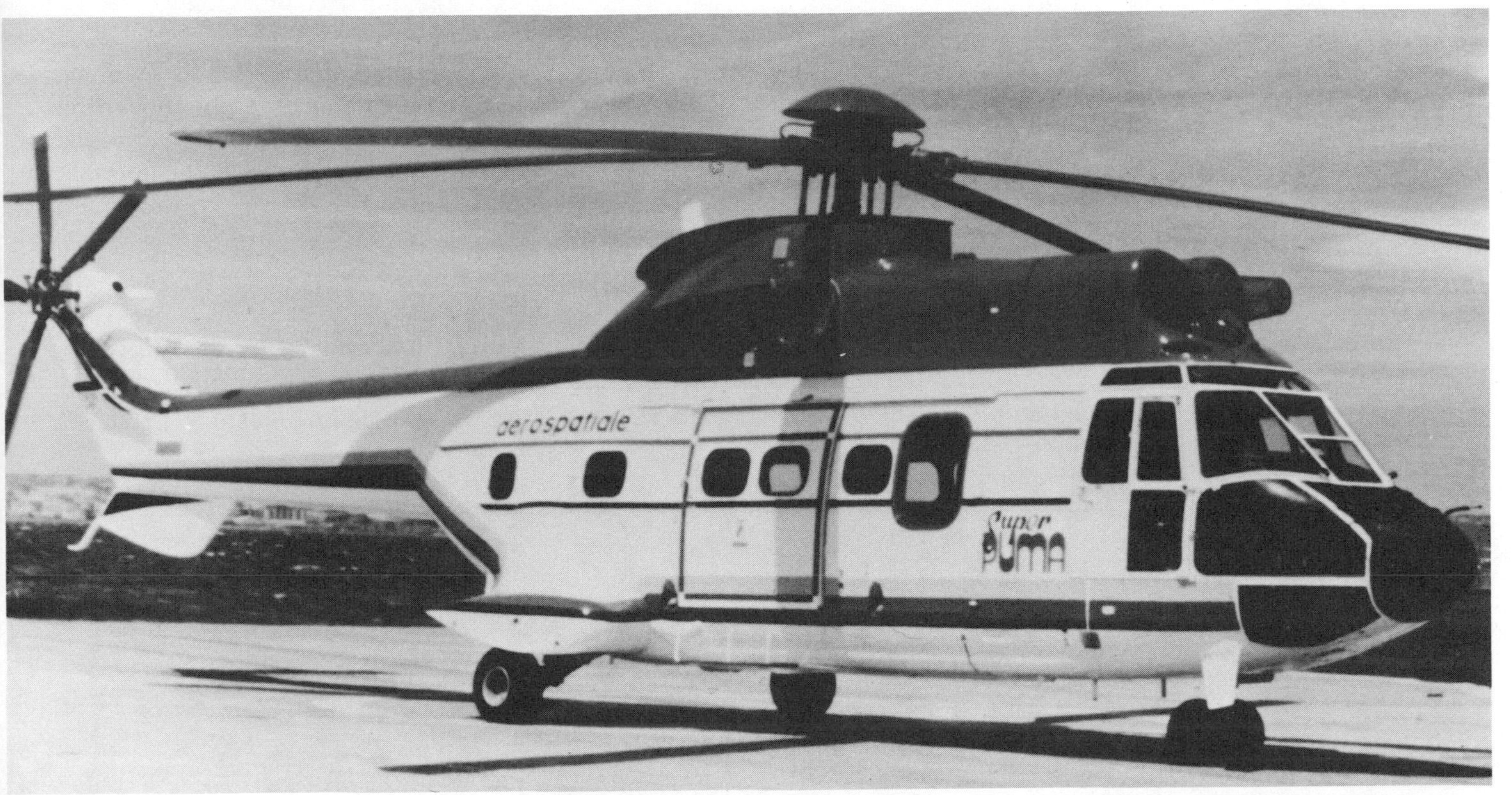

Aerospatiale AS 332C Super Puma

Manufacturer and Model	Seats	Powerplant	Fuel (gal, no rsv)	Weights Gross/ Empty (lbs)	Cruise Speed (mph)	Never Exceed Speed (Vne)	Max Range m @ alt	Hover OGE	Hover IGE
AERO-SPATIALE AS 332C Super Puma	23	2 Turbomeca Makila 1755 shp. ea.	408	17,200/ 8643	161 @ sea level	192	520 @ sea level	7,550′	9,840′

AS 355 TwinStar

The TwinStar's cabin can accommodate six people comfortably, and there is more than 35 cubic feet of baggage space in three separate compartments with individual access doors. Two GM Allison 2500-20F turbine engines power the 355 TwinStar, offering 420 shp each for takeoff and 370 shp each for normal cruising power.

Manufacturer and Model	Seats	Powerplant	Fuel (gal, no rsv)	Weights Gross/ Empty (lbs)	Cruise Speed (mph)	Never Exceed Speed (Vne)	Max Range m @ alt	Hover OGE	Hover IGE
AERO-SPATIALE AS 355 TwinStar	6	2 Allison 250C-20F 370 shp. ea.	190	4630/ 2741	137 @ sea level	169	499 @ sea level	4,920′	7,220′

Aerospatiale AS 355 TwinStar

109A

The 109A helicopter includes an aluminum alloy airframe and bonded panel fuselage, two doors on the right, two doors on the left, and a tinted acrylic windshield and six side windows with sliding ventilation panels on the pilot's right-side window. Two Allison model 250-C-208B turboshaft engines power the copter.

Agusta 109A

Manufacturer and Model	Seats	Powerplant	Fuel (gal, no rsv)	Weights Gross/ Empty (lbs)	Cruise Speed (mph)	Never Exceed Speed (Vne)	Max Range m @ alt	Hover OGE	Hover IGE
AGUSTA 109A	8	2 Allison 250-C-20B 420 shp ea.	146	5730/ 3636	172 @ 8,000′	193	472 @ 8,000′	6,800′	10,000′

Mark II

The Mark II resembles the 109A in appearance and is a modification of that helicopter.

BELL

206B JetRanger III

The five-place turbine JetRanger III serves many functions: speedy air taxi, executive transport, police aircraft, aerial crane, and all-around utility vehicle. Cabin interiors can be decorated in a variety of fabrics and colors. Special features include an autopilot and single-pilot IFR.

Bell 206B JetRanger III

Manufacturer and Model	Seats	Powerplant	Fuel (gal, no rsv)	Weights Gross/ Empty (lbs)	Cruise Speed (mph)	Never Exceed Speed (Vne)	Max Range m @ alt	Hover OGE	Hover IGE
BELL 206B Jet-Ranger III	5	Allison 250-C20B 420 shp.	76	3200/ 1580	136 @ sea level	150	345 @ sea level	8,800′	12,400′

Bell LongRanger II

206L-I LongRanger II

The seven-place, 140 mph turbine-powered LongRanger can fill a wide variety of single-engine helicopter requirements.

Manufacturer and Model	Seats	Powerplant	Fuel (gal, no rsv)	Weights Gross/ Empty (lbs)	Cruise Speed (mph)	Never Exceed Speed (Vne)	Max Range m @ alt	Hover OGE	Hover IGE
BELL 206L-1 Long-Ranger II	7	Allison 250-C28B 500 shp.	98	4050/ 2156	130 @ sea level	150	345 @ sea level	8,100′	13,200′

Bell 205A

205A

Bell's 15-place 205A is a design offspring of the company's Huey military helicopter, used for transporting people, hauling cargo, or as a two-ton aerial crane. The helicopter features sliding doors for easy entrance and exit, picture windows for viewing pleasure, and a sound-conditioned cabin for more comfort. Luggage can be transported in a separate compartment.

Manufacturer and Model	Seats	Powerplant	Fuel (gal, no rsv)	Weights Gross/ Empty (lbs)	Cruise Speed (mph)	Never Exceed Speed (Vne)	Max Range m @ alt	Hover OGE	Hover IGE
BELL 205A-1	15	Lyc. 1531-13B 1400 shp.	215	9500/ 5323	126 @ sea level	126	310 @ sea level	6,000′	10,400′

Model 222 can seat as many as 10 people in a variety of cabin interiors and with an optional climate control system.

222

Manufacturer and Model	Seats	Powerplant	Fuel (gal, no rsv)	Weights Gross/ Empty (lbs)	Cruise Speed (mph)	Never Exceed Speed (Vne)	Max Range m @ alt	Hover OGE	Hover IGE
BELL 222	8-10	2 Lyc. LTS-101-650C-2 615 shp ea.	184	7650/ 4550	165 @ sea level	184	395 @ sea level	6,400′	10,300′

Bell 222

Bell 212

212

The 212 was designed with twin powerplants, hydraulic systems, electrical systems, fuel systems, and even twin fire detectors and extinguishing systems. It is also IFR-certified to fly at night and in inclement weather, and is used in executive transport, construction work, exploration, offshore crew change, border patrol, and search-and-rescue operations. The 212 is available as a wide-body corporate executive transport seating from six to 10 persons and also in the 15-place general purpose configuration.

Manufacturer and Model	Seats	Powerplant	Fuel (gal, no rsv)	Weights Gross/ Empty (lbs)	Cruise Speed (mph)	Never Exceed Speed (Vne)	Max Range m @ alt	Hover OGE	Hover IGE
BELL 212 Twin	15	2 P&W PT6T-3 900 shp. ea.	215	11,200/ 5933	115 @ sea level	115	260 @ sea level	9300′ @ 10,000 lbs.	11,000′

The 19-place 214ST is Bell's most powerful single turbine-powered aircraft, providing more lift and speed capabilities than any of its other utility helicopters. With quick release cargo hook, the 214 can lift over 7,000 pounds at sea level and move the cargo at 120 mph. IFR is part of the standard equipment.

214ST

Manufacturer and Model	*Seats*	*Powerplant*	*Fuel (gal, no rsv)*	*Weights Gross/ Empty (lbs)*	*Cruise Speed (mph)*	*Never Exceed Speed (Vne)*	*Max Range m @ alt*	*Hover OGE*	*Hover IGE*
BELL 214ST	18	2-G.E. CT 7-2 1625 shp. ea.	410	17,200/ 9500	NA	NA	480	3,300′	12,600′

NA–Not Available.

Bell 214ST

412

Production of the 412 began in January 1981, and it then became the first four-bladed production helicopter manufactured by Bell. Featuring improved twin PT6T-3B-1 Pratt & Whitney turbines, the 412 is designed for a 5,000-pound useful load capacity. The use of elastomeric bearings and dampers eliminate the need for lubrication in the rotor system.

Manufacturer and Model	*Seats*	*Powerplant*	*Fuel (gal, no rsv)*	*Weights Gross/ Empty (lbs)*	*Cruise Speed (mph)*	*Never Exceed Speed (Vne)*	*Max Range m @ alt*	*Hover OGE*	*Hover IGE*
BELL 412	15	2 P&W PT6T-3B 900 shp. ea.	215	11,600/ 6223	141 @ sea level	161	275 @ sea level	7,100′	11,000′

Bell 412

Bell 214B

The 214B, also known as Bell's Big Lifter, can seat as many as 16 persons. Engine takeoff power is 2,930; maximum takeoff/landing is 8,000; useful load is 7,700 pounds.

214B

Manufacturer and Model	Seats	Powerplant	Fuel (gal, no rsv)	Weights Gross/ Empty (lbs)	Cruise Speed (mph)	Never Exceed Speed (Vne)	Max Range m @ alt	Hover OGE	Hover IGE
BELL 214B Big Lifter	16	Lyc. T550-8D 2930 shp.	204	13,800/ 7827	161 @ sea level	161	198 @ sea level	10,400′	15,000′

BOEING VERTOL

Chinook 234

The 234 is a versatile helicopter suitable for passengers and cargo. As a cargo carrier, the helicopter features more than 1,449 cubic feet of usable volume in the wide, constant cross-section fuselage and an additional 227 cubic feet in the area over the ramp. Loads of up to 14 tons can be lifted on the external cargo hooks of the helicopter.

As a passenger carrier, the helicopter cruises at 140 knots, has a comfortable interior seating up to 44 passengers traveling as far as 600 miles, and includes underseat stowage for carry-on luggage. Larger items can be stored in the main luggage compartment aft of the passenger cabin. A refreshment galley and lavatory are included.

Manufacturer and Model	Seats	Powerplant	Fuel (gal, no rsv)	Weights Gross/ Empty (lbs)	Cruise Speed (mph)	Never Exceed Speed (Vne)	Max Range m @ alt	Hover OGE	Hover IGE
BOEING VERTOL Chinook 234	45	2 Lyc. AL 5512 4075 shp.	2090	47,000/ 25,500	157 @ sea level	190	627 @ sea level	4,900′	15,000′

Boering Vertol Chinook Model 234

BRANTLY-HYNES

Brantly-Hynes B-2 Series

B-2 Series

The B-2 series has a top speed of 100 mph, cruises at 90 mph, has a 250-mile range with reserve and gives its passengers roomy comfort. The 28-foot-long helicopter seats one crew member and one passenger.

Manufacturer and Model	Seats	Powerplant	Fuel (gal, no rsv)	Weights Gross/ Empty (lbs)	Cruise Speed (mph)	Never Exceed Speed (Vne)	Max Range m @ alt	Hover OGE	Hover IGE
BRANTLY-HYNES B2B	2	Lyc. IVO-360-A1A 180 hp	31	1670/ 1000	95 @ sea level	100	240 @ sea level	4,000′	6,700′

Brantly-Hynes B-2 Series

ENSTROM

F-28C-2

Powered by a Lycoming HIO-360-EIAD four-cylinder fuel-injected engine rated at 205 HP, the F-28C-2 model can accommodate three persons. The baggage compartment can house up to 100 pounds. Night lighting equipment that is standard on this aircraft includes two beacons and four running lights with internally lit instrument cluster, eyebrow panel lights, and panel light dimmer control. Options include cargo hooks, floats, sirens, and searchlights.

Manufacturer and Model	Seats	Powerplant	Fuel (gal, no rsv)	Weights Gross/ Empty (lbs)	Cruise Speed (mph)	Never Exceed Speed (Vne)	Max Range m @ alt	Hover OGE	Hover IGE
ENSTROM F-28C-2	3	Lycoming HIO-360-EIAD 205 hp.	40	2350/ 1500	85 @ sea level	112	270 @ sea level	8,300′	13,000′

Enstrom F-28C-2

Enstrom 280C Shark

280C Shark

The idler on the transmission of the 280C Shark allows passengers to board or deplane safely with the engine running and with rotors stopped. The helicopter performs a wide variety of missions such as emergency parts courier, site inspection, patrols, news coverage, and carrying of external loads. Its cruise speed is 95 mph (117 maximum) and its rate of climb is 1,150 fpm. The useful load is 850 pounds.

Manufacturer and Model	Seats	Powerplant	Fuel (gal, no rsv)	Weights Gross/ Empty (lbs)	Cruise Speed (mph)	Never Exceed Speed (Vne)	Max Range m @ alt	Hover OGE	Hover IGE
ENSTROM 280C Shark	3	Lycoming HIO-360-EIAD 205 hp.	40	2350/ 1500	95 @ sea level	117	300 @ sea level	8,300′	13,000′

F-28F

The F-28F is a functionally designed, three-place, turbocharged utility and high visibility helicopter with unlimited operations. Powered by a Lycoming four-cylinder injected engine rated at 225 HP, the helicopter can climb to a maximum approved operating ceiling of 10,000 feet.

Manufacturer and Model	Seats	Powerplant	Fuel (gal, no rsv)	Weights Gross/ Empty (lbs)	Cruise Speed (mph)	Never Exceed Speed (Vne)	Max Range m @ alt	Hover OGE	Hover IGE
ENSTROM F-28F	3	Lycoming HIO-360-F1AD 225 hp.	42	2350/ 1500	96 @ sea level	112	263 @ sea level	7,500′	12,400′

Enstrom F-28F

280F Shark

The 280F Shark is a three-place, turbocharged business and pleasure helicopter with unlimited operations. Powered by a Lycoming advanced HIO-360-FIAD, four-cylinder fuel-injected engine rated at 225 HP, the helicopter can cruise at 101 mph.

Manufacturer and Model	Seats	Powerplant	Fuel (gal, no rsv)	Weights Gross/ Empty (lbs)	Cruise Speed (mph)	Never Exceed Speed (Vne)	Max Range m @ alt	Hover OGE	Hover IGE
ENSTROM 280F Shark	3	Lycoming HIO-360-FIAD 225 hp.	42	2350/ 1500	101 @ sea level	117	277 @ sea level	7,500′	12,400′

UH12E

The UH12E includes many items as standard equipment, some of which are dual carburetors and high dome piston kit, separate transmission oil system, edge-lighted instrument panel, tinted glass, sliding window doors, ground handling wheels, eight-day clock, and seat belts for three occupants. Some of the accessories that are available are quick release cargo hook, radio, cabin heater and defroster, night lighting equipment, first aid kit, fire extinguisher, floor mats, and skid gear extensions for amphibious gear.

Hiller UH12E

Manufacturer and Model	Seats	Powerplant	Fuel (gal, no rsv)	Weights Gross/ Empty (lbs)	Cruise Speed (mph)	Never Exceed Speed (Vne)	Max Range m @ alt	Hover OGE	Hover IGE
HILLER UH12E	3	Lyc. VO-540-C2A 305 hp	46	3100/ 1759	90 @ sea level	95	214 @ sea level	6,800′	10,400′

UH12E4

The UH12E4 functions in many roles, some of which include an ambulance, agricultural sprayer, and emergency medical transport, for photography and as ski patrol. The helicopter can accommodate four occupants, including the pilot, in a roomy cabin with seat belts, ashtrays, eight-day clock, and battery temperature gauge all included in the standard equipment items.

Manufacturer and Model	Seats	Powerplant	Fuel (gal, no rsv)	Weights Gross/ Empty (lbs)	Cruise Speed (mph)	Never Exceed Speed (Vne)	Max Range m @ alt	Hover OGE	Hover IGE
HILLER UH12E4	4	Lyc. VO-540-C2A 305 hp.	46	3100/ 1836	90 @ sea level	94	214 @ sea level	6,800′	10,400′

Hiller UH12E4

Hiller UH12ET

UH12ET

The three-place turbine UH12ET has a maximum permissible speed of 96 mph, a cruise speed of 90 mph, and a maximum weight of 3,100 pounds. The engine, which weighs 158 pounds, is an Allison model 250-C20B turbine free turboshaft. The nonretractable skid landing gear allows for easy cabin access. Some of the accessories that are available include strobe lights, loudspeaker, amphibious landing gear, and cargo racks.

Manufacturer and Model	Seats	Powerplant	Fuel (gal, no rsv)	Weights Gross/ Empty (lbs)	Cruise Speed (mph)	Never Exceed Speed (Vne)	Max Range m @ alt	Hover OGE	Hover IGE
HILLER UH12ET (Turbine)	3	Allison 250-C20B 301 hp.	46	3100/ 1650	90 @ sea level	95	351 @ sea level	7,000′	12,000′

Hiller UH12E4T

UH12E4T

The UH12E4T is powered by the same engine as the UH12ET but can accommodate four persons. This helicopter, like the other Hiller models, comes in a choice of six colors with coordinated fabric upholstery and vinyl trim.

Manufacturer and Model	Seats	Powerplant	Fuel (gal, no rsv)	Weights Gross/ Empty (lbs)	Cruise Speed (mph)	Never Exceed Speed (Vne)	Max Range m @ alt	Hover OGE	Hover IGE
HILLER UH12E4T (Turbine)	4	Allison 250-C20B 301 hp.	46	3100/ 1650	90 @ sea level	95	351 @ sea level	7,000′	12,000′

UH FH 1100

A five-place turbine helicopter, the FH 1100 has standard features such as transmission oil pressure gauge, tinted glass, cargo hook hard point, ground handling wheels, sliding windows, seat belts for all occupants, and custom polyurethane paint scheme.

NOTE: Specifications refer to model Fairchild produced. Hiller's new version will have a higher horsepower engine and a new main rotor blade.

Manufacturer and Model	*Seats*	*Powerplant*	*Fuel (gal, no rsv)*	*Weights Gross/ Empty (lbs)*	*Cruise Speed (mph)*	*Never Exceed Speed (Vne)*	*Max Range m @ alt*	*Hover OGE*	*Hover IGE*
HILLER (FAIRCHILD) UH FH 1100	5	Allison 250-C11B	68/ 136	2750/ 1395	142 @ 4,000′	128	396 @ 5,000′	8,400′	13,400′

Hiller FH 1100

Hughes 300C

300C

The 300C is a versatile, piston-powered helicopter used for executive transport, as a pilot trainer, as a construction lifter, for patrol, and as an aerial applicator. It can fly up to 15,000 feet, cruise at speeds to 100 mph, has a range to 225 miles on normal fuel and 380 miles with its auxiliary tank and uses only 10 gallons of fuel per hour.

Manufacturer and Model	Seats	Powerplant	Fuel (gal, no rsv)	Weights Gross/ Empty (lbs)	Cruise Speed (mph)	Never Exceed Speed (Vne)	Max Range m @ alt	Hover OGE	Hover IGE
HUGHES 300C	3	Lyc. HIO-360-D1A 225 hp.	30	2050/ 1046	95 @ 4,000′	104	225 @ 4,000′	2,750′	5,900′

Hughes 500D

500D

The 500D light turbine helicopter allows easy configuration conversions from a deluxe five-place executive transport to a utility cargo carrier or a short-haul vehicle carrying pilot and six passengers. Typical uses include intracity and intercity transportation of key personnel; aerial survey, patrol, and photography; agricultural duties such as seeding, spraying, and fire fighting; air rescue operations; and numerous tasks connected with the construction, petroleum, and forestry industries. The aft compartment can carry two 55-gallon drums on a flat floor. Outsize bulky loads can be carried externally by a cargo hook, which is rated at over one ton.

Manufacturer and Model	Seats	Powerplant	Fuel (gal, no rsv)	Weights Gross/ Empty (lbs)	Cruise Speed (mph)	Never Exceed Speed (Vne)	Max Range m @ alt	Hover OGE	Hover IGE
HUGHES 500D	5-7	Allison 250-C20B 420 shp.	64	3000/ 1360	160 @ 5,000′	157	329 @ 5,000′	7,500′	8,500′

MBB

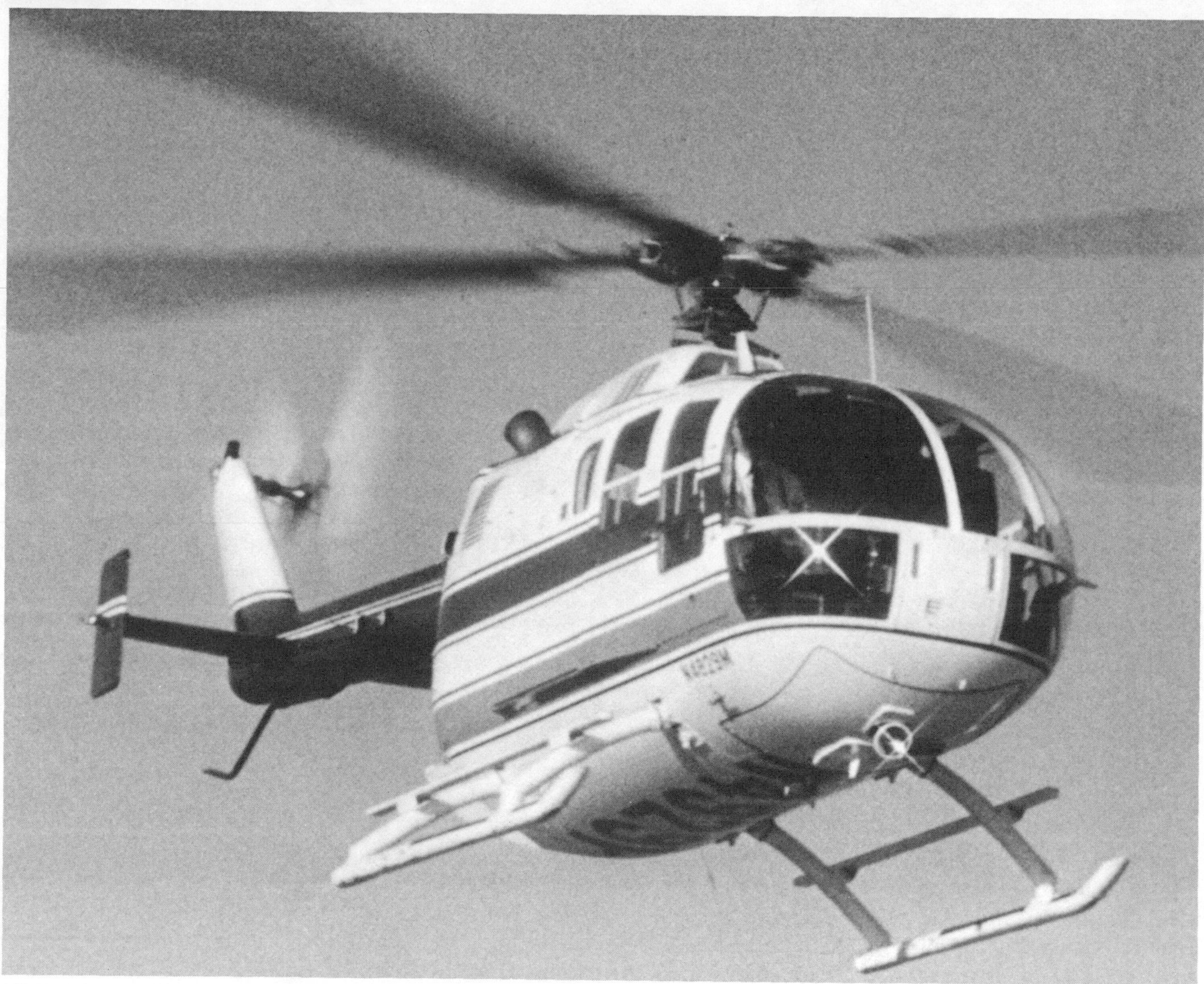

MBB BO 105 CBS

BO 105 CBS

For the BO 105 CBS, there are two Allison 250-C20B engines, a useful load of 2,616 pounds, a standard fuel capacity of 1,014 pounds, and a rate of climb at sea level of 1,476 feet/min. Main optional equipment includes windshield wiper, heating system, main rotor blade folding, engine fire extinguishing system, emergency floats, snow skids, settling protectors, and cargo hook.

Manufacturer and Model	Seats	Powerplant	Fuel (gal, no rsv)	Weights Gross/ Empty (lbs)	Cruise Speed (mph)	Never Exceed Speed (Vne)	Max Range m @ alt	Hover OGE	Hover IGE
MBB BO 105 CBS	NA	2 Allison 250-C20B 380 shp. ea.	1720 lbs.	5291/ 2675	150 @ sea level	167	357 @ sea level	5,300′	8,400′

NA–Not Available

The BK 117 helicopter has a powerplant of two Lycoming LTS 101-650-B-1 turboshaft engines. Main optional equipment includes windshield wiper for the co-pilot, bleed air heating system, emergency floats, settling protectors, snow skids, and dual controls.

BK 117

Manufacturer and Model	Seats	Powerplant	Fuel (gal, no rsv)	Weights Gross/ Empty (lbs)	Cruise Speed (mph)	Never Exceed Speed (Vne)	Max Range m @ alt	Hover OGE	Hover IGE
MBB BK-117	NA	2 Lyc. LTS-101-650 493 shp. ea.	1764 lbs.	6173/ 3395	164 @ sea level	170	338 @ sea level	9,975′	13,190′

NA–Not Available

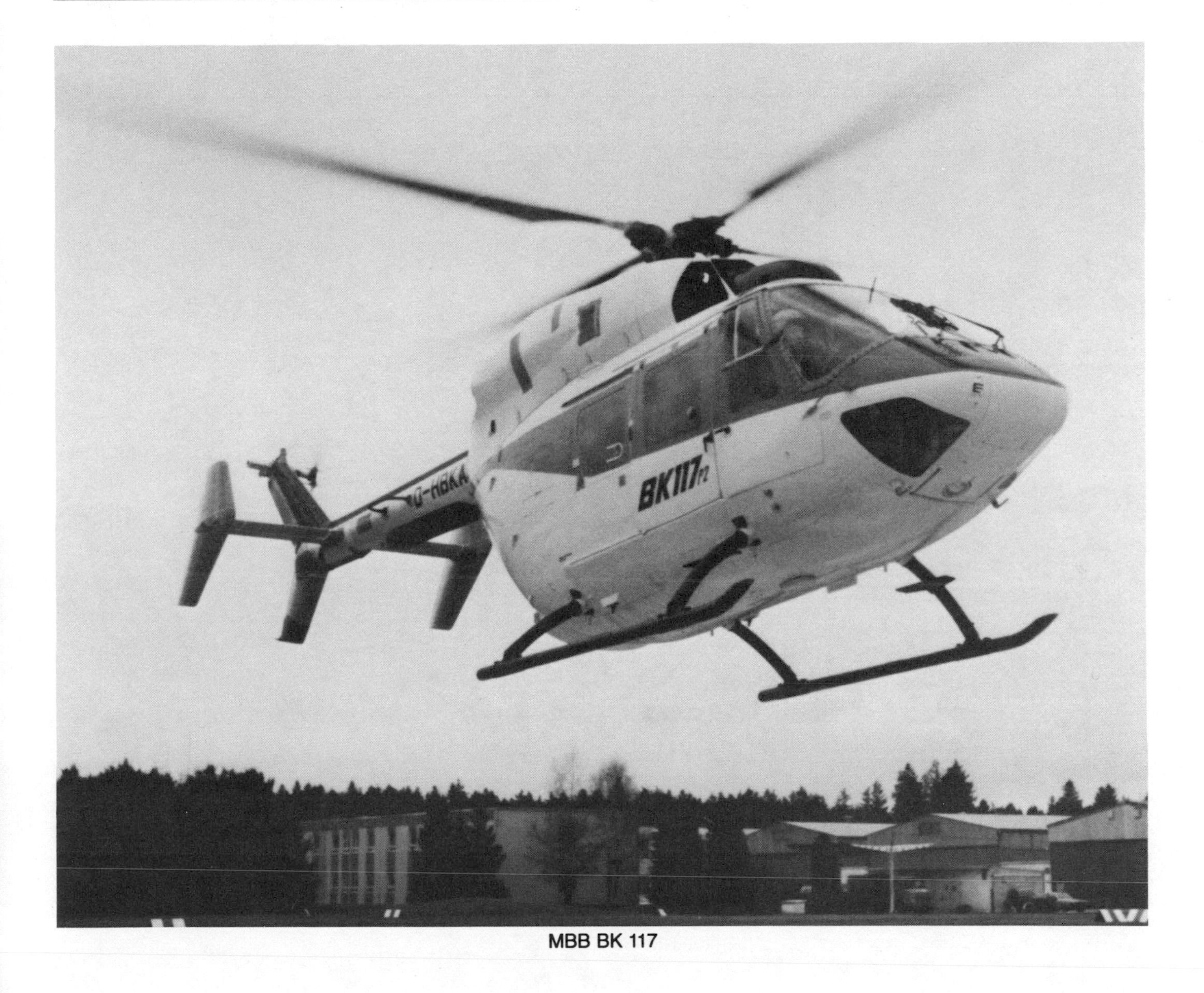

MBB BK 117

ROBINSON

R-22

The R-22 is an ultralight, two-place helicopter with standard equipment of dual controls and a small area for baggage under both seats. Lights for night flying and a variety of avionics are offered as optional equipment. The starting procedure for the helicopter has been simplified. The engine is started by turning an automotive style ignition switch, and then is allowed to idle for 30 or 40 seconds while the clutch engages automatically. The instrument scan is assisted by warning lights near the critical gauges. After a quick mag and sprag unit check, the helicopter is ready for flight.

Manufacturer and Model	Seats	Powerplant	Fuel (gal, no rsv)	Weights Gross/ Empty (lbs)	Cruise Speed (mph)	Never Exceed Speed (Vne)	Max Range m @ alt	Hover OGE	Hover IGE
R-22	2	Lyc. 0–320 124 hp.	20	1300/ 796	108 @ sea level	118	240 @ sea level	4,500′	6,500′

Robinson R-22

Exec

A fully enclosed, streamlined, teardrop fuselage gives the Exec its high speed performance. The use of a water-cooled engine makes this possible. The Exec airframe features a combination of space frame and monocoque construction.

Rotorway EXEC

Manufacturer and Model	Seats	Powerplant	Fuel (gal, no rsv)	Weights Gross/ Empty (lbs)	Cruise Speed (mph)	Never Exceed Speed (Vne)	Max Range m @ alt	Hover OGE	Hover IGE
ROTORWAY Exec	2	RW-145 Water Cooled 4 stroke 145 hp.	14/19	1263/ 780	105 @ sea level	125	234 @ sea level	4,000′	9,500′

Scorpion 133

Lightweight for easy transport and compact in size, the Scorpion 133 fits comfortably in most garages alongside a car. A simple system of three belts in tandem drives the Scorpion's tail rotor. The V-belt system absorbs the vibrations of the rotor, eliminates the need for a tail rotor gearbox and offers a tremendous cost savings over other designs. The powerplant is a four-cylinder, four-stroke, horizontally opposed, water-cooled aircraft engine.

Manufacturer and Model	*Seats*	*Powerplant*	*Fuel (gal, no rsv)*	*Weights Gross/ Empty (lbs)*	*Cruise Speed (mph)*	*Never Exceed Speed (Vne)*	*Max Range m @ alt*	*Hover OGE*	*Hover IGE*
ROTORWAY Scorpion 133	NA	NA	NA	NA	80 @ sea level	NA	120 @ sea level	NA	5,500′

NA–Not Available

Rotorway Scorpion 133

S-76

The S-76 is a twin-engine helicopter with four-blade main rotor with titanium spars, fiberglass skin and aftswept tips; retractable tricycle landing gear; 45-foot-long fuselage with four large cabin doors; and full instrumentation, with provisions for the installation of IFR equipment. The S-76 can also be fitted with a variety of optional equipment: extended range fuel tanks, air conditioning, weather radar, rescue hoist, cargo hook, emergency equipment, and engine air particle separators.

Sikorsky S-76

Manufacturer and Model	Seats	Powerplant	Fuel (gal, no rsv)	Weights Gross/ Empty (lbs)	Cruise Speed (mph)	Never Exceed Speed (Vne)	Max Range m @ alt	Hover OGE	Hover IGE
SIKORSKY S-76 (Spirit)	14	2 Allison 250-C30 1300 shp. ea.	280	10,000/ 5475	166 @ sea level	178	553 @ 3,000′	2,900′	6,200′

Sikorsky ABC

ABC

The ABC (Advanced Blade Concept) technology demonstrator aircraft was designed and built to evaluate the feasibility of the ATC rotor system through flight testing as a pure helicopter and at high speed in the auxiliary propulsion mode using side-mounted jet engines. The ABC system uses two counter-rotating rigid main rotors mounted one above the other on a common shaft. It combines the advantages of a low-speed helicopter with those of a high-speed aircraft without the need for a wing.

Specifications are not yet available since this helicopter is still an experimental model.

Bensen	Dept. PP-129 P.O. Box 31047 Raleigh, North Carolina 27612	B-80 B8M Powerplant Module Engines Options B-8 Gyroglider B-8 Hydrocopter
Champ	P.O. Box 21 Woodland Hills, California 91365	7AC Champ
Christen Industries, Inc.	1048 Santa Ana Valley Road Hollister, California 95023	Christen Eagle I Christen Eagle I-F Christen Eagle II
Great Lakes Aircraft	Box 3526 Enid, Oklahoma 73701	2T-1A-2
Ken Brock Manufacturing	11852 Western Avenue Stanton, California 90680	KB-2 Gyroplane Engines Catalog
Mudry Aviation	Route 376 Dutchess County Airport Wappingers Falls, New York 12590	Cap 10B
Pitts Aerobatics	P.O. Box 547 Afton, Wyoming 83110	S-1S S-2A
Trimcraft Aero	6254 Highway 36 Burlington, Wisconsin 63105	Custom Wood Kits
Weedhopper	Box 2253 Ogden, Utah 84404	Gypsy Weedhopper

BENSEN AIRCRAFT

Bensen Aircraft B-80

The Bensen B-80 kit contains all materials and hardware prefabricated by their factory as far as the FAA permits (less engine installation). The kit also includes all the special tools needed for the gyrocopter's fabrication and assembly. Also by building the plane oneself, the FAA permits owners to service their own plane. Endurance of the gyrocopter with a six-gallon tank is two hours. The range is 120 mph. Because the gyrocopter is small enough to fit into a home garage, owners can save on hangar fees. Average time to learn to fly it is three to five hours for beginners, two to three hours for licensed airplane pilots.

B-80

ADVISORY CIRCULAR

The following Federal Aviation Regulations and Advisory Circulars are pertinent to the construction and operation of amateur-built aircraft and may be obtained from the Superintendent of Documents, U. S. Government Printing Office, Washington, D. C. 20402, at the price indicated:

a. FAR, Volume II - contains Parts 11, 13, 15, 21,37, 39, 45, 47, 49, 183, 185, 187, and 189 — $6.00
b. FAR, Volume VI - contains Parts 91, 93, 99, 101, 103, and 105 — 5.00
c. FAR, Volume IX - contains 61, 63, 65, 67, 141, 143, and 147 — 6.00
d. Advisory Circular No. 43.13-1 — 3.00
e. Advisory Circular No. 43.13-2 — 2.00

The above includes amendment transmittal service .

WILLIAM G. SHREVE, JR.
Acting Director
Flight Standards Service

B-8M

The construction plans for the B-8M model show exactly what and how the gyrocopter should be built, give specifications, include detailed drawings, tools needed, and choice of design variations. The 21-foot, two-blade rotor will not stall, slow down, or reverse rotation, even if its air speed is reduced to zero or if its engine quits. If the engine stops, the craft can glide down gently, like a parachute, and be landed without a bump. Powered by a 72 HP McCulloch engine, the B-8M cruises at speeds from 45 to 80 miles per hour and carries a useful load of 250 pounds, including pilot.

B-8M

POWERPLANT MODULE

This complete powerplant installation package contains all necessary components for the installation and use of the McCulloch 4318 engine (72 to 90 HP). It includes the newest style engine mount, fuel system, carburetor, jet cooling cowls, smooth short stroke throttle, all hardware and the Bensen 495 propeller. New, simplified, step-by-step fabrication and assembly instructions and manuals make it an easy engine installation package.

ENGINES

McCulloch engines are available from several sources. Prices may vary from one supplier to the next.

OPTIONS

Several options as to landing gear, powerplant, and enclosures have been tested by Bensen. The plans and drawings on how to construct and use these optional components are available. Some of the items necessary for construction are available in kit form and are listed in the Custom Conversion section.

NOTE: Bensen factory services everything it sells. All recent buyers of Bensen kits (within the last 12 months) receive "Bensen News" newsletters to keep them abreast of the latest developments.

B-8 Gyroglider

The B-8 Gyroglider is a simplified version of the copter without the engine. The glider, either on wheels or on pontoons, flies when towed aloft by a car or a boat, or can soar like a kite in a strong wind. It can be converted into a gyrocopter by adding an engine.

B-8 Hydrocopter

The B-8 hydrocopter is the standard B-8M gyrocopter with floats installed.

Write manufacturer for complete details and prices.

7AC Champ

The 7AC Champ is a fully FAA-certificated design that is easy to build, fuel efficient, and low in cost. It can be built in a home garage or basement. The plane has a welded tubular steel fuselage frame and tail, spring steel landing gear legs, formed aluminum wing ribs, quality wooden spars and lifetime covering. Both basic and deluxe kits are available.

Manufacturer and Model	Seats	Powerplant /Propeller	Fuel Capacity (std/opt. gal)	Weights Gross/ Empty (lbs)	Cruise Speed (mph) 75% @ alt	Optimum Range (m w 45 min. rsv) 75% @ alt	Takeoff/ Landing Distance (ft. over 50′ obst)	Rate of Climb (fpm)	Service Ceiling	Stall Speed (gear/flaps down, mph)
CHAMP 7AC Champ	2	Perf. based on 60 hp engine	14	1220/ 770	86 @ 4,000	300/NA	900/NA	460	NA	40

NA–Not Available

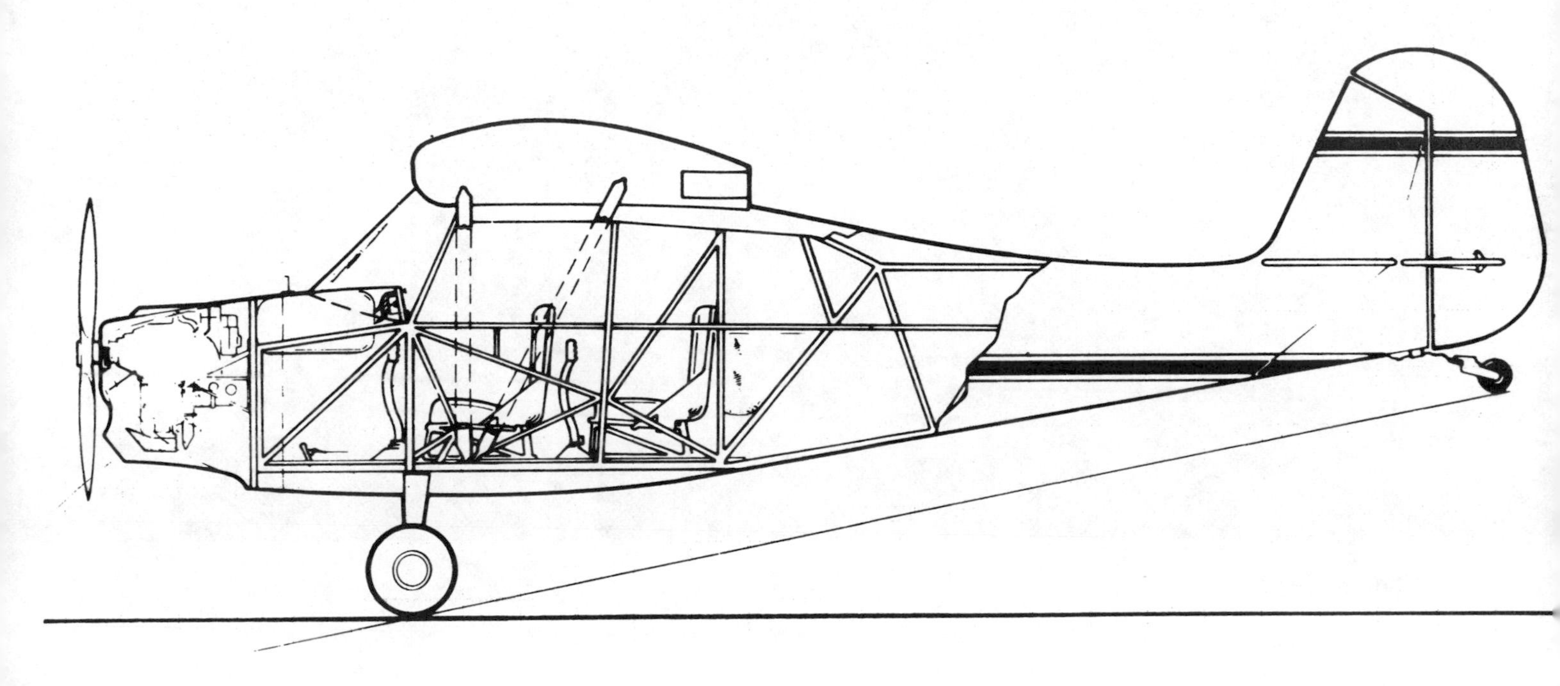

CHRISTEN INDUSTRIES

Christen Eagle I

The Christen Eagle I is a special-purpose single-place biplane intended for unlimited-class aerobatic competition. It is lightweight and equipped with removable auxiliary fuel tank. The Eagle I uses a constant-speed propeller.

Manufacturer and Model	Seats	Powerplant /Propeller	Fuel Capacity (std/opt. gal)	Weights Gross/ Empty (lbs)	Cruise Speed (mph) 75% @ alt	Optimum Range (m w 45 min. rsv) 75% @ alt	Takeoff/ Landing Distance (ft. over 50′ obst)	Rate of Climb (fpm)	Service Ceiling	Stall Speed (gear/flaps down, mph)
CHRISTEN INDUSTRIES Christen Eagle I	1	Lyc. AEIO 540-D485- 260 hp.	25 (Aux 20)	1478/ 997	171	350/NA or 635/NA	2100/ NA	2640	NA	60

NA–Not Available.

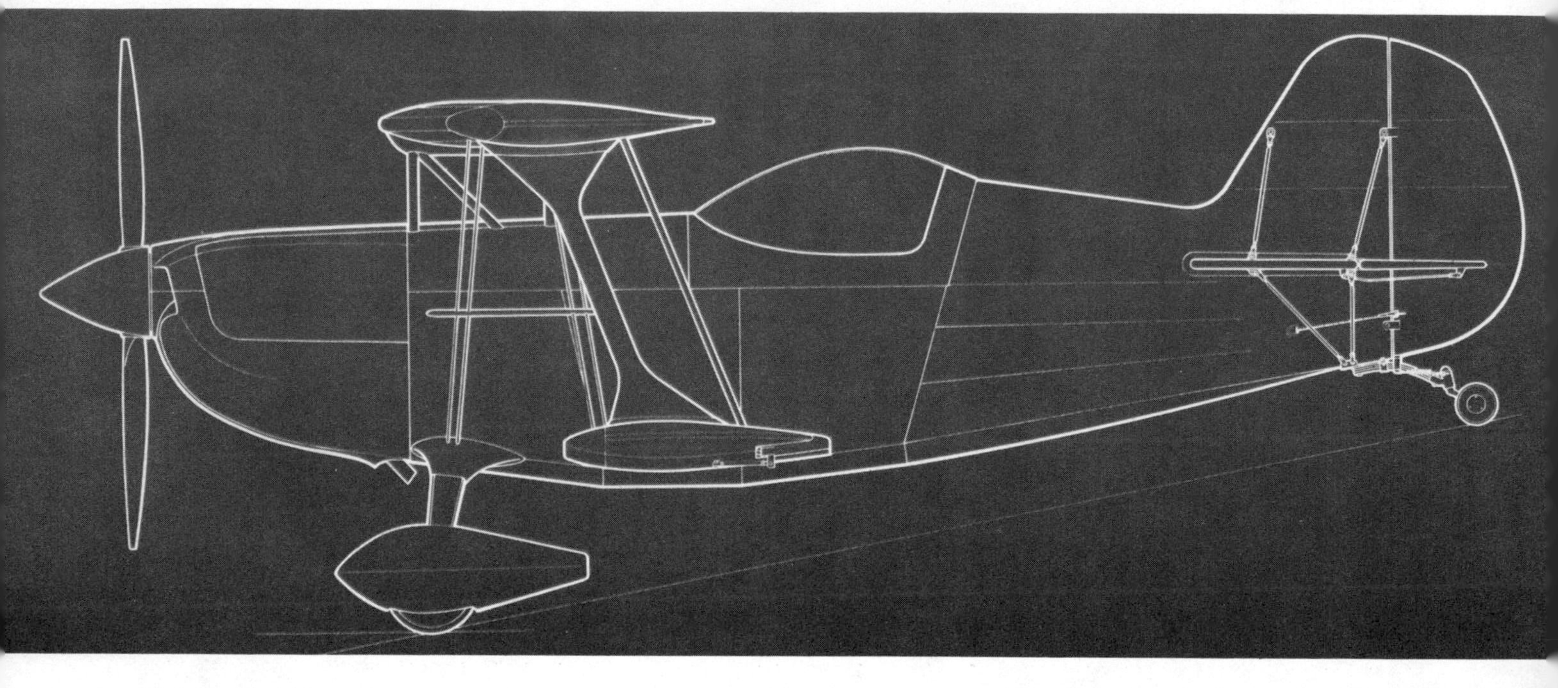

Christen Eagle I

Christen Eagle I-F

The Christen Eagle I-F is a special-purpose single-place biplane intended for unlimited-class aerobatic competition. It is lightweight and equipped with removable auxiliary fuel tank. The Eagle I-F uses a fixed-pitch propeller.

Manufacturer and Model	*Seats*	*Powerplant /Propeller*	*Fuel Capacity (std/opt. gal)*	*Weights Gross/ Empty (lbs)*	*Cruise Speed (mph) 75% @ alt*	*Optimum Range (m w 45 min. rsv) 75% @ alt*	*Takeoff/ Landing Distance (ft. over 50' obst)*	*Rate of Climb (fpm)*	*Service Ceiling*	*Stall Speed (gear/flaps down, mph)*
CHRISTEN INDUSTRIES Christen Eagle II	1	Lyc. AEIO 540-D4B5 266 hp.	25 (Aux 20)	1478/ 978	162	325 @ NA or 602 @ NA	1950/NA	2600	NA	60

NA–Not Available.

Christen Eagle I-F

Christen Eagle II

The Eagle II is a two-place sport biplane, fully capable of unlimited-class aerobatic competition, but designed to provide additional pilot/passenger comfort and convenience. This model features a full electrical system (starter-alternator-battery), and uses a 200 HP engine with constant-speed propeller.

Manufacturer and Model	Seats	Powerplant /Propeller	Fuel Capacity (std/opt. gal)	Weights Gross/ Empty (lbs)	Cruise Speed (mph) 75% @ alt	Optimum Range (m w 45 min. rsv) 75% @ alt	Takeoff/ Landing Distance (ft. over 50′ obst)	Rate of Climb (fpm)	Service Ceiling	Stall Speed (gear/flaps down, mph)
CHRISTEN INDUSTRIES Christen Eagle II	1	Lyc. AEIO 360-AIO 200 hp.	25	1578/ 1025	165	380 @ NA	1500/ NA	2100	NA	58

NA–Not Available.

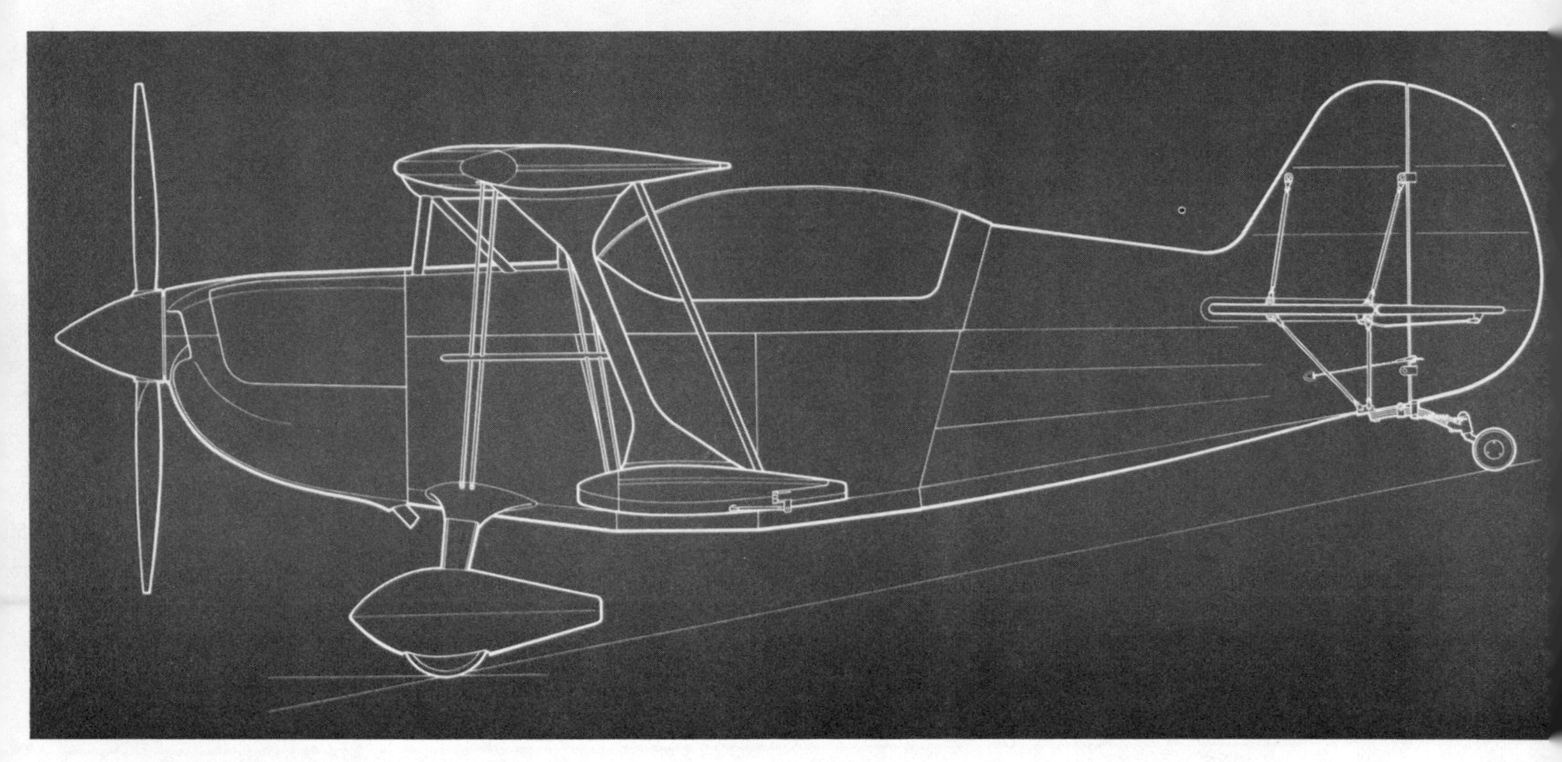

Christen Eagle II

Great Lakes Aircraft 2T-1A-2

2T-1A-2

The sporty, two-place 2T-1A-2 airplane features a 180 HP Lycoming engine, inverted oil and fuel system, four ailerons instead of two for good maneuverability, and navigation lights for early dawn or late evening flights. Approved aerobatic maneuvers include spins, chandelles, lazy eights, loops, barrel rolls, point rolls, slow rolls, snap rolls, primary rolls, hammerhead turn, immelmann, split "S" and Cuban 8's.

Manufacturer and Model	*Seats*	*Powerplant*	*Fuel Capacity (std/opt. gal)*	*Weights Gross/ Empty (lbs)*	*Cruise Speed (mph) 75% @ alt 65% @ alt*	*Optimum Range (m w 45 min. rsv) 75% @ alt 65% @ alt*	*Takeoff/ Landing Distance (ft. over 50′ obst)*	*Rate of Climb (fpm)*	*Service Ceiling*	*Stall Speed (gear/flaps down, mph)*
GREAT LAKES AIRCRAFT 2T-1A-2 Biplane (Aerobatic)	2	Lyc. AEIO-360-B1G6 180 hp.	NA	1800/ 1230	118 @ NA NA	375 @ NA NA	825/ 825	1150	17,000′	NA

NA–Not Available.

KEN BROCK MANUFACTURING

KB-2 Gyroplane

This gyroplane is constructed using eight available sub-kits provided by Ken Brock Mfg. The builder has the option of purchasing the complete package all at once or each sub-kit separately.

KB-2 Gyroplane

ENGINES

Ken Brock Mfg. offers either the McCulloch (72 or 90 HP) or the VS engine converted to aircraft use to its customers along with the appropriate engine installation kit.

CATALOG

Ken Brock Mfg. also offers a catalog listing gyroplane accessories such as propellers, fuel lines and fittings, goggles, and tachometers. In addition, parts for the various aircraft are listed with photographs. To obtain a copy, write for current price to Ken Brock Mfg., Inc., 11852 Western Avenue, Stanton, California 90680.

MUDRY AVIATION

Cap 10 B

The Cap 10 is an all wood tail wheel design, with dual sticks and throttles. Its 180 HP Lycoming engine is fuel injected with a full Christen inverted system. The Cap 10 can serve commercial uses: for the beginning student it is a fine basic trainer; for the advanced pilot it is a perfect cross-country aircraft with 156 mph cruise speed. Its low operating and maintenance cost makes the Cap 10 a good sport aircraft. Because of its slow landing speed, rugged landing gear, and ground maneuvering characteristics, it can be flown in and out of the smallest airports in this country.

Manufacturer and Model	Seats	Powerplant	Fuel Capacity (std/opt. gal)	Weights Gross/ Empty (lbs)	Cruise Speed (mph) 75% @ alt 65% @ alt	Optimum Range (m w 45 min. rsv) 75% @ alt 65% @ alt	Takeoff/ Landing Distance (ft. over 50′ obst)	Rate of Climb (fpm)	Service Ceiling	Stall Speed (gear/flaps down, mph)
CAP 10 B (Aerobatic)	2	Lyc. AEIO 360-B2F 180 hp.	40	NA	NA	NA	NA	NA	NA	NA
Cap 21 (Aerobatic)	2	Lyc. AEIO 360-A1B 200 hp.	19	1322/ 1080	167 @ NA	NA	NA	NA	NA	52

NA–Not Available

Mudry CAP 10 B

PITTS AEROBATICS

S-1S

The S-1S model can be purchased as a complete airplane ready to fly or can be assembled from a kit. The engine is the rugged fuel-injected Lycoming AEIO-360-B4A with a stronger crankshaft, stronger crankcase, and stronger cylinder bases than earlier models. The engine is equipped with inverted lubrication and fuel system to improve the aerobatic capability and durability of the engine when flown aerobatically. The propeller is fixed pitch metal with FAA aerobatic approval.

Manufacturer and Model	Seats	Powerplant	Fuel Capacity (std/opt. gal)	Weights Gross/ Empty (lbs)	Cruise Speed (mph) 75% @ alt 65% @ alt	Optimum Range (m w 45 min. rsv) 75% @ alt 65% @ alt	Takeoff/ Landing Distance (ft. over 50′ obst)	Rate of Climb (fpm)	Service Ceiling	Stall Speed (gear/flaps down, mph)
PITTS S-1S (Biplane) (Aerobatic)	1	Lyc. AEIO-360-B4A 180 hp.	20	1150/ 750	145 @ NA 135 @ NA	230 @ NA NA	1050/ 970	2600	22,000′	57

NA–Not Available.

Pitts Aerobatics S-1S

The S-2 was designed with the same goals in mind as the S-1S, but with the ability to carry two people. It is fully certified by the Federal Aviation Administration in the aerobatic category and can be purchased as a complete plane ready to fly. It is capable of doing all the maneuvers in the Aresti system. The plane includes all of the latest design improvements, including the patented Pitts symmetrical wing system. The engine is the rugged fuel-injected Lycoming AEIO-360-AIA, which is equipped with an inverted lubrication and fuel system to improve its aerobatic capability and durability of the engine when flown aerobatically. The propeller used is the Hartzell constant speed metal propeller, which is approved for unlimited aerobatics with the AEIO-360-AIA engine.

S-2A

Manufacturer and Model	Seats	Powerplant	Fuel Capacity (std/opt. gal)	Weights Gross/ Empty (lbs)	Cruise Speed (mph) 75% @ alt 65% @ alt	Optimum Range (m w 45 min. rsv) 75% @ alt 65% @ alt	Takeoff/ Landing Distance (ft. over 50' obst)	Rate of Climb (fpm)	Service Ceiling	Stall Speed (gear/flaps down, mph)
PITTS S-2A (Biplane) (Aerobatic)	2	Lyc. AEIO-360-A1A 200 hp	24	1500/ 1035	147 @ NA 139 @ NA	230 @ NA NA	1275/ 1230	1950	22,000'	57

NA–Not Available.

Pitts Aerobatics S-2A

TRIMCRAFT AERO

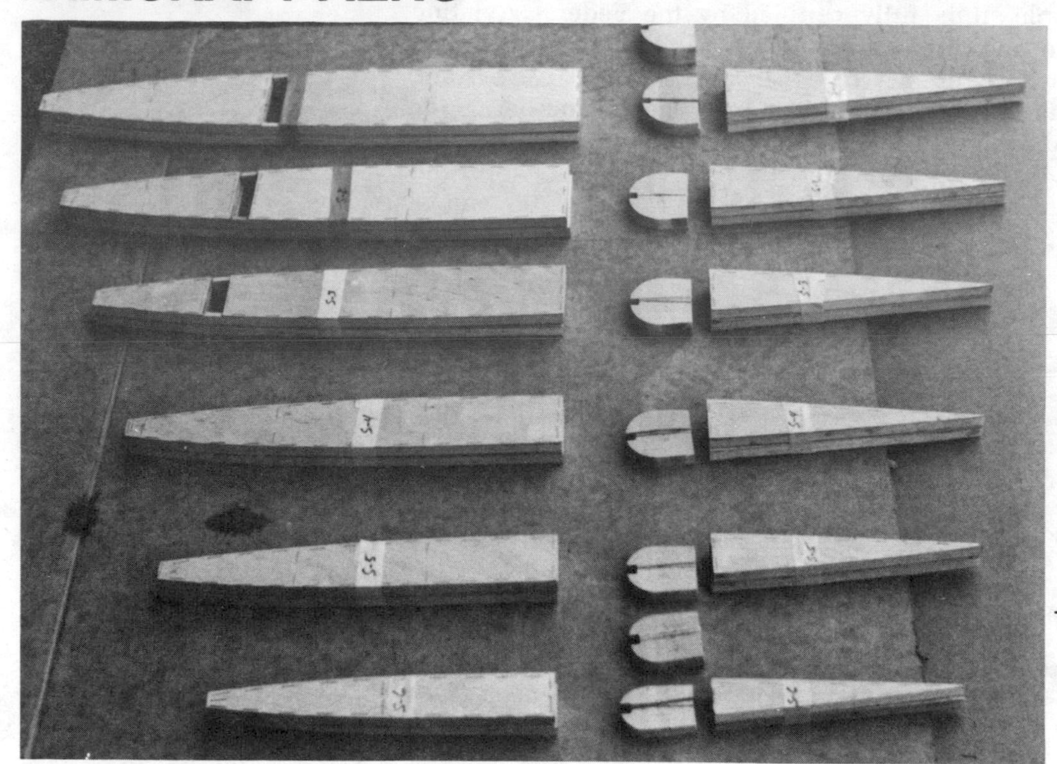

Trimcraft Aero Custom-Milled Wood Homebuilt Aircraft Kits

Custom-milled Wood Homebuilt Aircraft Kits

The milled spar kits are for those persons who require only the spars. Each kit contains all wing, aileron, and center section spars where required. Spars are beveled and tapered. The spars in spar kits are not drilled. The milled kits contain all parts milled to proper shape and size as much as possible without doing any assembly work. These kits can be assembled with tools found in the average home shop. All kits take approximately 25 to 30 days.

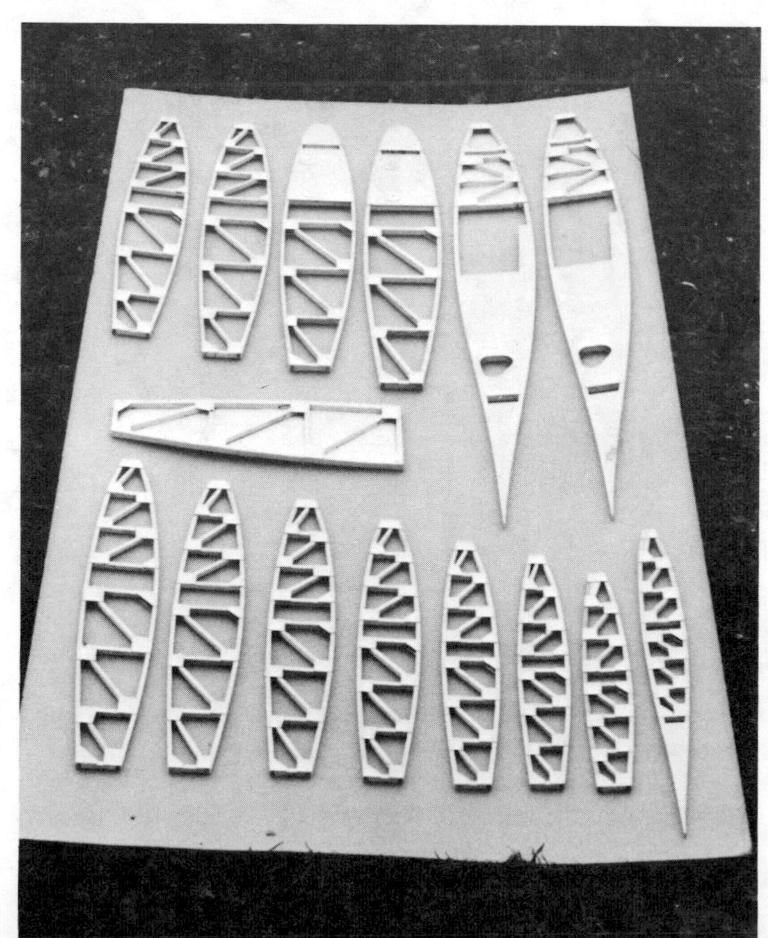

Trimcraft Aero Custom-Milled Wood Homebuilt Aircraft Kits

WEEDHOPPER

Gypsy

The Gypsy, also sold in kit form by Weedhopper, utilizes an aluminum frame styrofoam ribs and fuselage and Chotia-460 engine which develops 25 HP and weighs only 32 pounds. Takeoff roll at sea level with an average-size pilot is well below 100 feet and can be executed on sod strips, gravel roads, or smooth fields. Rate of climb is about 500 fpm. Landing rollout will use up about 250 feet of pavement.

Weedhopper Gypsy

Manufacturer and Model	Seats	Powerplant	Fuel Capacity (std/opt. gal)	Weights Gross/ Empty (lbs)	Cruise Speed (mph)	Optimum Range	Takeoff/ Landing Distance (ft. over 50′ obst)	Rate of Climb (fpm)	Service Ceiling	Stall Speed (mph)
WEED-HOPPER Gypsy Ultra Light	1	Chotia-460 25 hp.	2	415/ 165	35-40	120	100/250	500	12,000′	22

Weedhopper

The Chotia-460 engine powers the Weedhopper, which is built from a kit in about 40 hours. The standard 28-foot wings are easy to handle in moderate turbulence. The long-span 34-foot wings, an extra cost option, increase soaring ability. The two different wings can be changed in the field in less than half an hour and both pack up for easy transport on top of a car. The covering is stabilized dacron sailcloth, pre-sewn to slip in place. Because the Weedhopper can be launched and landed on the pilot's feet, no license is required for the plane or pilot.

NOTE: A two-place Weedhopper Two with a 34-foot-span soaring sail will be available to order as of March 1, 1982.

Manufacturer and Model	Seats	Powerplant	Fuel Capacity (std/opt. gal)	Weights Gross/ Empty (lbs)	Cruise Speed (mph)	Optimum Range	Takeoff/ Landing Distance (ft. over 50′ obst)	Rate of Climb (fpm)	Service Ceiling	Stall Speed (mph)
WEED-HOPPER Weedhopper Ultra Light	1	456CL Chotia, 2 cy., 25 hp.	3.5	380/ 160	30	90	100/ NA	600	10,000′	22

NA–Not Available.

Weedhopper

Many companies specialize in modifying and converting existing aircraft in order to improve the airplane's performance, give it better climbing power, faster cruising speeds, greater fuel efficiency, or to adapt it for specialized needs.

For example, the Garrett Corporation can convert a noisy and fuel-guzzling HS 125, 1A through 400 Series, to a quiet and fuel-efficient 731/HS 125 by retrofitting Garrett TFE 731 turbofan engines. The modification takes less than 90 days.

The following companies offer some of the most extensive modification services in the field.

AiRESEARCH AVIATION

AiResearch Aviation offers a complete range of services for the business aviation owner: engine and airframe maintenance, avionics installation and servicing, interior completions and refurbishing, re-engining and airframe modifications, product support, exterior painting, line and customer services.

Garrett, AiResearch Aviation Company
6201 West Imperial Highway
Los Angeles, California 90045

ATLANTIC AVIATION

Atlantic owns more than 200 Supplemental Type Certificates issued by the FAA for modifications on more than 15 types of fixed and rotary wing aircraft. The company also offers an aircraft maintenance and inspection center; avionics design, installation and repair with more than 335 navigation and communications systems; interior completion and refurbishment; aircraft finishes, design, and application including custom paint schemes; aircraft replacement parts.

Atlantic Aviation Corporation
Greater Wilmington Airport
P.O. Box 15000
Wilmington, Delaware 19850

AYRES CORPORATION

Ayres offers firewall-forward conversion kits, enabling operators to switch to the Pezetel PZL-3S engine and Vibra-Damp mount. The PZL-3S engine has seven cylinders, single row radial, air-cooled, direct propeller drive with counter-clockwise rotation (viewed from rear), single-speed supercharger.

Ayres Corporation
P.O. Box 3090
Albany, Georgia 31706

CAPRE

The Aqua Float from Capre becomes an integral part of an airplane. Features include a flat top, solid keels, lift boosters, anodized extrusions, improved

Aqua Float Model 1500 from Capre on Taylor Craft Plane

Aqua Float Model 180-85 from Capre on Cessna 180 plane

double with rudders, rubber nose bumpers, flat skin aluminum construction, and seven compartments in each float.

Capre
805 Geiger Road
Zephyrhills, Florida 33599

COLEMILL

The Beech Baron can be converted to Colemill's President 600. The modification consists of installing 300 HP Continental engines, three-bladed Hartzell propellers, and Woodward governors. Also part of the conversion are vacuum pumps, tac generators, rebuilt alternator or generator, new fuel and oil lines, and baffling overhaul. There is no cowling modification necessary and the performance is increased approximately 20 mph at cruise speed and single engine rate of climb is nearly double. All models will have a new gross weight of 5,100 pounds.

The Cessna 310 can be converted to the Executive 600. This conversion is accomplished using new Continental 10-520 engines, three-bladed McCauley propellers, and new Woodward governors. There is no cowling modification necessary and the only instrument changes are tachometer and fuel flow meter. The speed is 15 to 20 mph faster than the standard 310 with greatly increased climb. The single engine ceiling is increased approximately 4,000 feet.

The Navajo can be converted to a Panther Navajo with factory-new 350 HP turbocharged TIO 540 J2BD engines, new Hartzell four-bladed propellers, heavy-duty brakes, and wing tip landing lights.

Colemill Enterprises
P.O. Box 60627
Nashville, Tennessee 37206

Colemill "Zip Tip"

Cessna TU206 and Cessna 185 on PK D3500A Amphibious Floats

DEE HOWARD COMPANY

The Dee Howard Company offers many modification services:

Complete demate inspection, including replacing hydraulic and fuel hoses; removing and replacing foam insulation in required areas; inspecting and remating wings and fuselage, horizontal stabilizer, windshields, control surfaces, and landing gear; performing X ray, magnaflux, and dye penetration checks in required areas; performing corrosion treatment and repainting as necessary.

Application of thrust reversers on Learjets, which add safer operation, reduced maintenance costs due to longer brake and tire life, reduced insurance premiums and an increase in the value of the aircraft.

Installation of the Howard Spar Mod on Beechcraft that have the weldon kit installed, providing a permanent fix.

Installation of a certified Garrett Environmental Control Unit system for the Bell 206B JetRanger, 206L, and L-1 LongRanger helicopters, offering inflight and ground cabin cooling as well as cabin heating and humidity control; three-position bleed valve reduces bleed air penalty when only partial heating or cooling is required; routing or air ducts and installation of 14 individual outlets for cabin air distribution for uniform heating and cooling.

Dee Howard Company
P.O. Box 17300, International Airport
San Antonio, Texas 78217

DEVORE

The rugged construction of Devore's PK floats consists of heavy-gauge aluminum alloy sheet metal, extruded keel, chine, and deck coaming members, and closely spaced bulkheads and skin stiffeners. Each aluminum part is alodined and primed in detail for maximum corrosion resistance. Sealing compounds are cured in a temperature-humidity controlled oven for maximum water tightness.

Devore
6104B Kircher Boulevard N.E.
Albuquerque, New Mexico 87109

EDO

Many aircraft are licensed for use with Edo floats. A list showing the airplane company, model, engine specifications, and the corresponding float model is available upon request from Edo.

For example, Edo model 3430 floats can be fitted on the Cessna 206 series. Model 3430 floats have flat decks for secure footing in loading and unloading the aircraft, and bilge tubes for quick and easy draining of the floats in the event of seepage or condensation after long exposure. Other Edo 3430 float features include fluted bottom for maximum hydrodynamic performance, five watertight compartments, dual water rudders standard.

Edo
65 Ruchmore Street
Westbury, New York 11590

FIBERFLOAT

A new design of floats by Fiberfloat are made of Kevlar, honeycomb, and epoxy. Kevlar is many times stronger than fiberglass of the same weight and the final product is over five times stronger than steel by weight. Using an inverted "V" hull, the floats capture all of the water and force it under the float, creating lift, rather than suction (and also keep the spray off the prop) and cause the plane to ride higher in the water on takeoff run. It also provides for a lower lift-off speed.

Fiberfloat
895 East Gay Street
Bartow, Florida 33830

Cessna 206 with Edo model 3430 floats

Cessna 172 on Fiberfloat 3500

Maule M-5 on Fiberfloat 2400

Piper PA28-235 on Fli-Lite 2500 Skis

FLUIDYNE ENGINEERING

Fluidyne manufactures several types of winter skis for airplanes:

For the winter flyer who does not need to use cleared runways, Wheel Replacement Skis are an economical answer. Because the wheels are removed, no weight is added by skis and the plane's balance and center of gravity are near normal. The same is true for useful load and handling characteristics.

Airglide Combination Wheel Skis are controlled hydraulically from the cockpit, or manually operated on the ground. For cleared runways, the skis are raised and wheels are used normally. In ski position, the tires rest on the skis (which are lowered) and the shock absorbers and landing gear function as usual. The skis are made of an all-metal, rugged aluminum and steel construction. Hydraulic actuators are enclosed in the ski beam, protected from ice, slush, and snow. High performance plastic bottom laminates are standard for slush and sticky snow, resisting and minimizing freeze-down.

Fli-Lite Combination Wheel Skis are interchangeable side to side. For wheel operation, a plate under the tire is moved forward by hydraulic power and the ski is raised vertically for ample ground clearance. For ski operation, the ski is lowered and a plate slides under the wheel to keep it off the ground. Because the ski motion is essentially vertical, there is little shift in the center of gravity, so the airplane's balance and center of gravity remain near normal. Plastic bottom laminates are standard for minimum friction and to prevent freeze-down.

Fluidyne Engineering
5916 Olson Highway
Minneapolis, Minnesota 55422

RAM AIRCRAFT MODIFICATIONS

RAM modifies and/or remanufactures engines for increased horsepower as certified in new production aircraft. The company also provides custom refinishing and custom new interior installation. Two conversion kits for Cessna 400 series aircraft are also available.

RAM Aircraft Modifications
Waco-Madison Cooper Airport
P.O. Box 5219
Waco, Texas 76708

RILEY AIRCRAFT CORPORATION

Riley Aircraft Corporation upgrades business twins, providing them with more horsepower, greater performance and better safety margins.

Cessna 421 models are beefed up with the 600 HP Lycoming LTP 101 turbine along with a three-bladed 90-inch Hartzell propeller complete with auto-feather full BETA, reversing, and synchronizer. The Riley conversion of this aircraft is known as the Riley Jet Prop 421.

The Cessna 340 is also converted by Riley with a pair of Lycoming LTP 101 turboprops. It is known as the Riley Jet Prop 340.

Another Riley conversion, the Riley Rocket 340, has the Avco Lycoming TIO-540-R engine. In addition to the powerplant conversion, the Riley Rocket 340 includes Hartzell propellers.

The Riley Rocket Power 414 has 400 HP, eight-cylinder, direct-drive Lycoming 10-720 engines having dual Riley AiResearch turbocharger systems delivering 300 HP each to 33,000 feet. The Riley Rocket Power 414 cruises at 300 mph at 25,000 feet.

Three-blade Hartzell propellers provide advanced aerodynamic advantages for greater prop efficiency.

Riley Aircraft Corporation
2016 Palomar Airport Road
Carlsbad, California 92008

ROBERTSON AIRCRAFT CORPORATION

The modifications designed, manufactured and installed by Robertson Aircraft are aimed at improving the low-speed performance of single- and twin-engine general aviation aircraft to increase safety during the critical takeoff and landing portions of every flight. Modifications include:

Robertson-equipped Super Seneca II has spoilers instead of ailerons to give it more positive roll control, less chance of spins, no adverse yaw.

Robertson-equipped Bonanza AT-1 also has spoilers instead of ailerons.

Cessna 172 with 220HP Franklin Engine Conversion Kit

Robertson's slower takeoffs and approaches expand the 182RG's flight speed envelope at the slow end, and Cessna's retractable gear extends cruise and top speeds envelope at the high end.

Robertson's new and improved high-differential drooped-aileron system was designed for high performance single-engine aircraft, such as the pressurized Cessna P210, 210, T210.

Robertson's R/S system can be incorporated on an Aztec to lower takeoff and landing speeds for safer operation, to provide dual engine-driven hydraulic system; give a good single-engine rate-of-climb speed closer to lift-off speed.

Robertson's Hi-Lift systems reduce takeoff distance over a 50-foot obstacle by 15 percent for the 310 Twins; reduce lift-off speed by 18 percent; reduce accelerate-stop distance by 1300 feet.

Robertson's Hi-Lift systems reduce takeoff distance over a 50-foot obstacle by 26 percent for the 340 Twins; reduce lift-off speed by 11 percent; reduce accelerate-stop distance by 1140 feet.

Robertson Aircraft Corporation
839 W. Perimeter Road
Renton Municipal Airport
Renton, Washington 98055

SEAPLANE FLYING

In addition to engine conversion kits, Seaplane also has a STOL kit for Cessna 170-B, 172s, and the 175 model. The kits take from two to eight days to install, depending on the shop doing the job.

Seaplane Flying
Pearson Air Park
P.O. Box 2164
Vancouver, Washington 98661

SEGUIN AVIATION

The primary business of Seguin Aviation is the modification of the Piper PA-23 Apache, the 235 Apache, and the Aztec into the Standard Geronimo.

Seguin Aviation
2075 Highway 46
Seguin, Texas 78155

UNIVAIR AIRCRAFT CORPORATION

The Manufacturing Division of Univair Aircraft Corporation specializes in the manufacture of thousands of parts for "out of production" aircraft, including the Stinson 108 series, the Swift, the Ercoupe, Forney, Alon, and Mooney M-10 for which the Type Certificates are held. Type Certificates are also held for the Aermatic and Flottorp lines of propellers, which are also manufactured under Production Certificate. Parts for many other "out of production" aircraft include the Cessna, Piper, Luscombe, plus others produced under FAA Parts Manufacturing Approval.

Univair Aircraft Corporation
Route 3, Box 59
Aurora, Colorado 80011

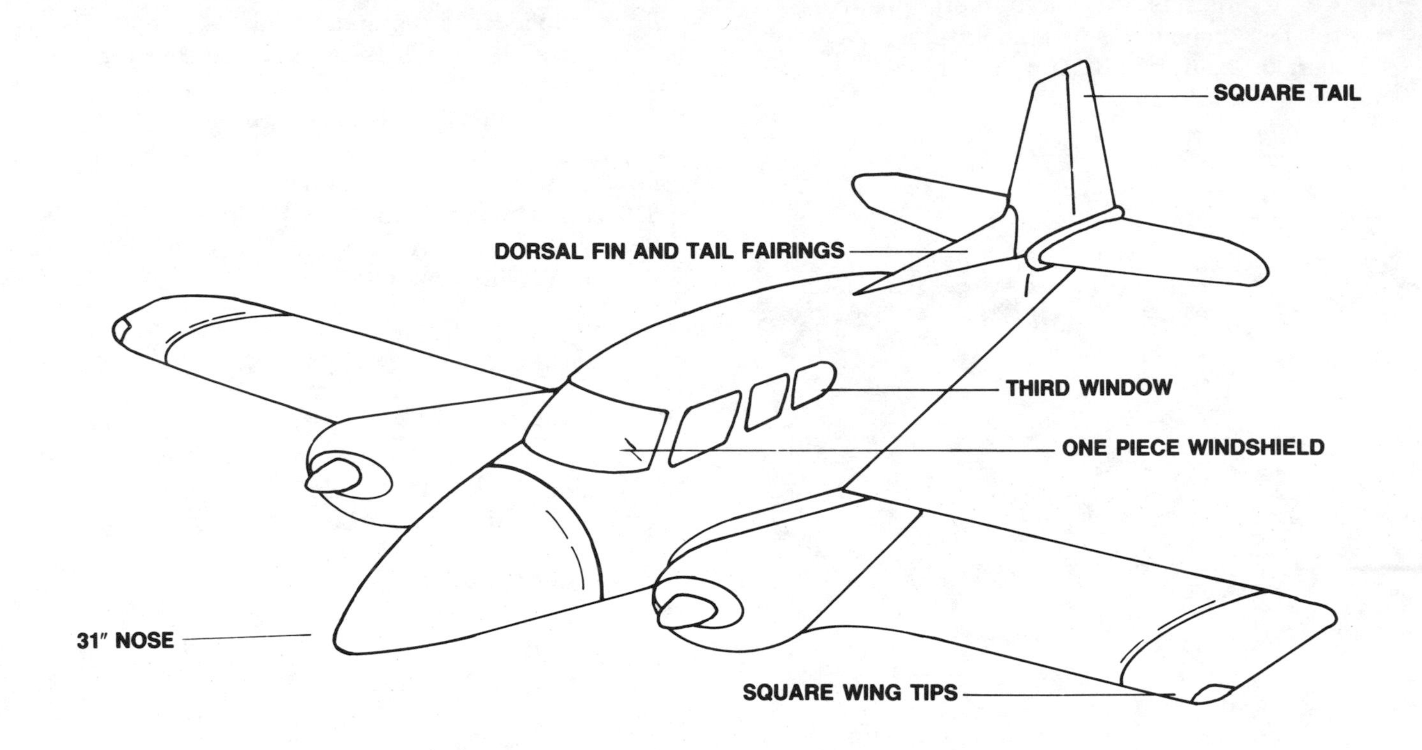

3 / Supplies for Maintaining an Aircraft's Engine, Interior, and Exterior

Now that you've purchased an airplane or are seriously contemplating the idea, you'll want to know how to keep both the exterior and interior of a plane in shape. This chapter is divided into three sections to do just that.

The first section provides information concerning an airplane's engine. The second section lists sources for material to decorate or improve the interior of an airplane. The third section provides information and sources for materials to improve or protect the exterior of an airplane.

ENGINE

AEROQUIP AEROSPACE DIVISION

Aeroquip manufactures rubber hose and fittings and nonrubber constructions, couplings, clamps, bellows, and conoseal joints.

Aeroquip Corporation
300 S. East Avenue
Jackson, Michigan 49203

AEROTECH PUBLICATIONS

Federal Aviation Regulations state that an aircraft owner/operator is responsible for the accuracy and completeness of maintenance records. Using the Adlog system, a pilot or owner can check the complete compliance, inspection, and service bulletin status as well as the routine maintenance history of his or her aircraft.

Each Adlog system combines maintenance logs with the complete text of all applicable airworthiness directives issued to an aircraft, its engine(s), propeller(s), rotors, and accessories.

The Adlog system is especially helpful when a person is buying or selling a used aircraft, in order to prove that a plane is or is not fully up to date as far as compliance and maintenance requirements are concerned.

For complete information, write to:

Aerotech Publications
P.O. Box 528
Old Bridge, New Jersey 08857

ALCOR, INC.

Alcor manufactures a variety of products:

Exhaust Gas Temperature (EGT) Mixture Control Indicators reduce maintenance costs through more accurate mixture control. A thermocouple placed in the exhaust pipe of a cylinder of the engine transmits its signal to a panel meter. As the pilot leans the mixture, the EGT rises as excess fuel is reduced. When no excess fuel remains, "peak EGT" is reached. Continued leaning from "peak EGT" causes a reduction in the EGT because of the cooling effect of the excess air ingested. The indicator also increases time between overhauls.

Exhaust Gas Temperature (EGT) Combustion Analyzer Systems monitor the exhaust gas temperature of every cylinder and allows the pilot to determine which cylinder is the leanest to enjoy optimum mixture control and most important to have a means for engine trouble detection. For carburetor engines, the cylinder-to-cylinder mixture distribution can vary considerably with altitude, outside air temperature, throttle setting. If the mixture is set to a cylinder not the leanest, the leaner cylinder(s) can be too lean. By changing throttle position slightly or adding alternate air, the mixture distribution can often be improved to allow better distribution to all cylinders and further reduce fuel consumption. Fuel injection engines can vary more than 200° F between cylinders.

The 50 amp Alternators (Model 1255a-12 volt) provide overvoltage relay protection for avionics equipment.

The Vernier Controls provide accuracy in mixture settings when an EGT Indicator is installed on small aircraft.

The EGT System Testers provide heat and voltage tests of EGT systems. The AlCal EGT System Testers permit testing any EGT system by applying a millivolt signal to simulate thermocouple output or by heating the exhaust probe with an electric heater to obtain temperatures up to 1,650° F.

The AlCal CHT System Testers eliminate errors in cylinder head temperature indicators, preventing engine failures.

TCP Concentrate eliminates spark plug fouling and associated lead problems from the antiknock additive, tetraethyl lead, that is found in aviation gasolines.

Alcor, Inc.
10130 Jones-Maltsberger Road
P.O. Box 32516
San Antonio, Texas 78284

CME, INC. Model 50

CME, INC.

From CME come the following ground test systems:

Vol-O-Flo meters consist of a laminar flow element and a differential pressure indicator or transducer.

Model 10 designates a flow element only, for use with a customer's P device.

Model 50 comes in two versions. 50-1 can be used with an inclined manometer (draft gauge) or with a differential pressure gauge of low range. It can also be used with certain differential pressure electrical transducers. 50-2 is a combined element and sensitive P gage.

FSC-1 is a portable system for making gas flow measurements over a wide range of flows at atmospheric pressures.

The Oxygen Test Stand tests Diluter Demand Oxygen Regulators under simulated altitude conditions.

There are other equipment items also available.

CME, Inc.
9431 Main Street
P.O. Box 1826
Manassas, Virginia 22110

KS AVIONICS

The Mixture-Mizer by KS Avionics monitors and analyzes exhaust gas temperature for both single and twin-engine planes. Pilots can adjust the mixture by responding to the exhaust gas temperature

(EGT) readout of a chromel-alumel thermocouple probe. The Mixture-Mizer's scale covers 250° F in 25° increments.

In addition, KS Avionics' Cylinder Head Thermalarm protects an airplane's engine from overheating by warning whenever a preset temperature limit is exceeded.

The Mixture-Mixer with Thermalarm warns of changes in operating conditions, descent with mixture leaned for cruise, and impending fuel exhaustion. Thermalarm warns of unwarranted rises in EGT through mixture change or other causes. The alarm triggers when a pointer passes a diamond symbol on the scale. In cruising flight, the pointer is set on the reference mark, 25°F below the diamond. The pilot then knows if the mixture drifts into an excessively lean condition. This can happen when a pilot descends by accident from altitude with mixture leaned for cruise.

Two probe models are available for KS Avionics, Mixture-Mizer EGT system: the Economy Probe and the Long-Life Probe. In choosing a probe, the following table can be followed:

PROBE MODEL	RECOMMENDED APPLICATIONS	WARRANTY
Economy, $20.00 Response Time, 0.8 sec.	Continental 0-360, Lycoming 0-320 and smaller engines (nonsupercharged)	1 year or 500 hours
Long-Life, $37.00 Response Time, 1.5 sec.	Continental 0-470, Lycoming 0-360 and larger engines as well as supercharged smaller engines.	2 years or 1000 hours

It should be pointed out that in some applications operators of engines as large as the Continental 0-470 will find the Economy probe adequate. The Economy probe may be evaluated at no additional cost, as a credit for its full price will be allowed in a later exchange for a Long-Life probe.

Equipment can be purchased as a complete system or as individual components.

KS Avionics
25216 Cypress Avenue
Hayward, California 94544

INTERIOR

AIRTEX PRODUCTS

Airtex provides owners of light planes with a source for new textile products they can install themselves. Items include upholstery sets and upholstered foam cushion sets, wall panel sets, headrest covers, carpets, baggage compartments, safety belts, fluorescent wind socks, protective covers for canopy, engine, propeller, windshields, and oil tanks, plus miscellaneous recovering supplies. Also available are aircraft finishes, including enamels, paints, and primers, nitrate dope, butyrate dope, and special finishes.

Airtex Products, Inc.
259 Lower Morrisville Road
Fallsington, Pennsylvania 19054

COMMONWEALTH CUSHIONS

Commonwealth Cushions distributes temper foam products that are available for a broad range of cushioning and protective padding applications in aircraft, medical, automotive, sports, and industrial products.

There are three types of temper foam cushions to choose from: standard temper foam cushions, multi-

KS AVIONICS Mixture-Mizer

AIRTEX PRODUCTS in 1965 Cessna 172

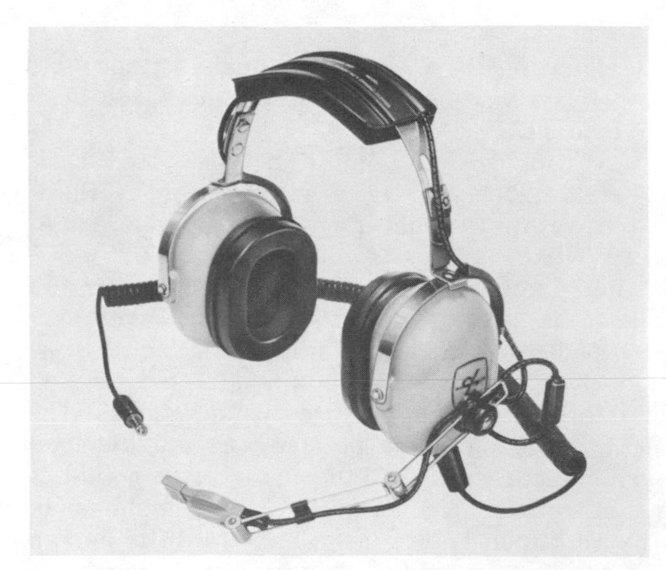
H10-76 Noise Attenuating Aviation Headset with Noise Cancelling Dynamic Microphone

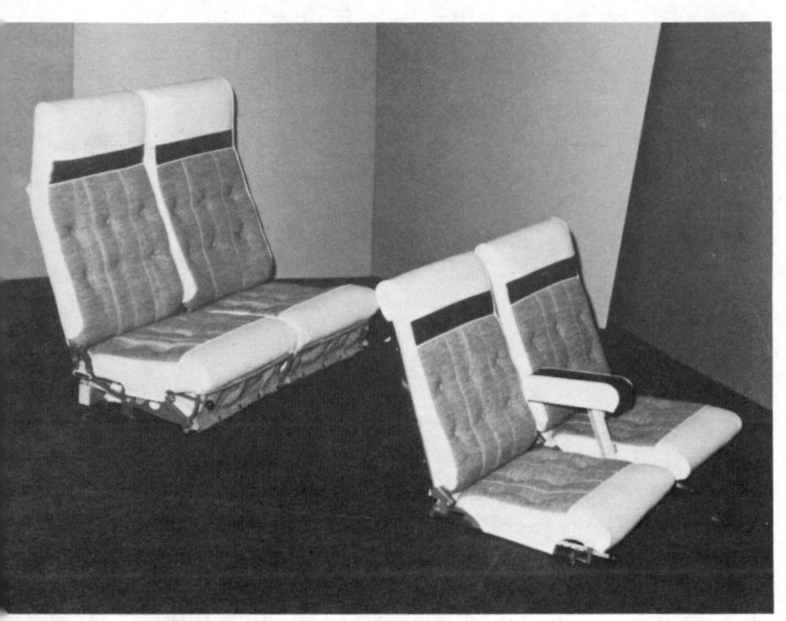
AIRTEX PRODUCTS in Beech K-35

layered temper foam cushions, and multilayered cushion with one-half-inch urethane top layer.

Commonwealth Cushions
P.O. Box 1084
Grafton Station
Yorktown, Virginia 23692

DALE & ASSOCIATES

The Emeco Division of Dale & Associates has engineered a high output cabin cooling system that is practical for light aircraft and can provide cooling breezes even on the ground during hot summer days. There are just three parts to the system: vent panel, screen, and blower.

Dale & Associates
Distributor Division
1401 Cranston Road
Beloit, Wisconsin 53511

JB SYSTEMS

JB manufactures air-conditioning systems for general aviation. The systems are operated electrically, from a pilot's internal power source while airborne, or from ground power supplies at the ramp. Electric drive allows ground cooling while loading, during preflight checks, or while waiting for takeoff. The systems can be retrofitted to a variety of airplanes. Write manufacturer for complete listing.

JB Systems
120 South Bowen Circle
P.O. Box 800
Longmont, Colorado 80501

MORGAN STANFORD AVIATION

Morgan Stanford manufactures the Thermacon-F Halogen heat screen and the Thermacon-MPA metallic screen. The aircraft interior and avionics are equally well protected by either the F Halogen or MPA screen.

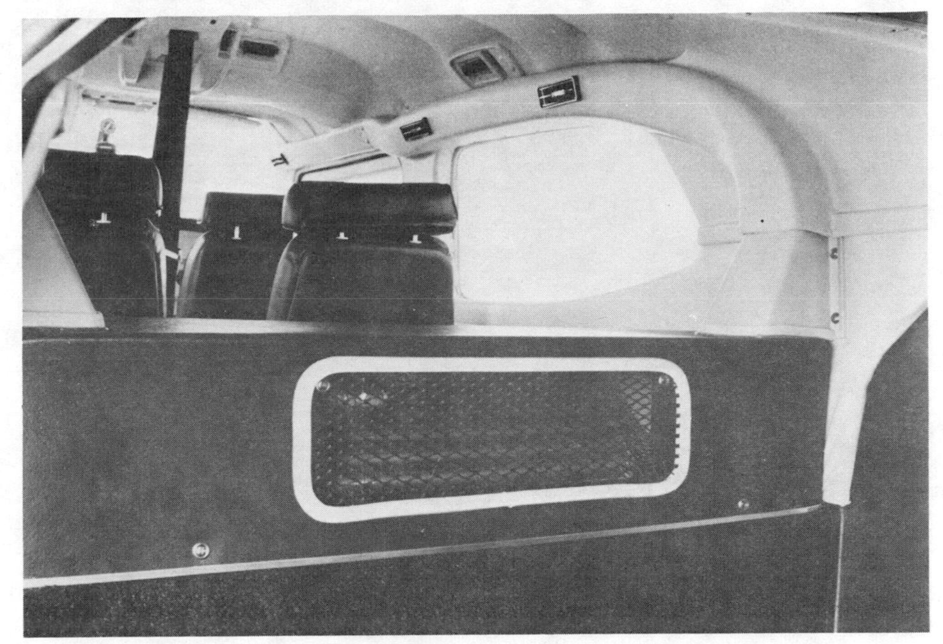

JB Air-Conditioning System

The F Halogen screen is recommended when a plane is never hangared and/or the owner expects to keep the same aircraft for many years. The MPA screen is best when the aircraft is frequently hangared and the owner does not expect to keep the same plane for long or when cost is a primary consideration.

Both screens can be attached or removed in seconds from over the interior of parked aircraft windows and will not scratch and craze windows.

Morgan Stanford Aviation
1805-J Second Street
Berkeley, California 94710

MONARCH TRADING CO. OF NEW ZEALAND

Monarch manufactures sheepskin seat covers that are easily fitted to the seat. They are also machine washable. The covers come in matched pairs to suit the two front seats of your aircraft. The five models made can be used for the Cessna 152; Cessna 172, 177, 180, and 182; Cessna 206, 207, and 210; Piper Tomahawk; Piper PA28 and PA32.

Write for current prices.

Monarch Trading Co. of New Zealand
Foreign Trade Zone, Pier 39
Honolulu, Hawaii 96817

Morgan Stanford Aviation Thermacon-F Halogen heat screen

Sheepskin Products Co. seat covers

SHEEPSKIN PRODUCTS CO.

Sheepskin seat covers protect expensive seats or cover worn places in older seats, and can be installed or removed in minutes. Colors: white, champagne, burnished gold, fawn, honey, oak, gray.

Sheepskin Products Co.
Box 733—803 Sherwood Drive
Richardson, Texas 75080

TANDY LEATHER COMPANY

Tandy manufactures ramskin seat covers that fit any size seat in cars, trucks, vans, and airplanes. The plush pile keeps pilot and passengers warm in winter and cool in summer. Covers include real sheepskin pelts in natural off-white; genuine sheepskin seat covers in camel and gray; two-tone seat covers of sheepskin in vertical or horizontal patterns in camel or gray; and custom sheepskin seat covers made to the exact specifications of the car or plane in gray, camel, or champagne.

Tandy Leather Company
P.O. Box 791
Fort Worth, Texas 76101

WAUGH ELECTRONICS

Waugh Electronics manufactures Skycaster, an aerial electronic billboard system consisting of a large grid of high-intensity light bulbs attached to the bottom of an airplane. Each bulb is individually programmed by a small, portable computer carried in the cockpit of the plane. This computer controls the advertising message as it travels across the sign from right to left. The sign and programming unit add only 76 pounds of weight to the aircraft.

The on-board programming unit is simple to operate so that a pilot can easily change messages in mid-flight.

The Skycaster can be installed on any single-engine airplane with a 50 amp or larger alternator—right down to a small plane such as a Cessna 150. And there are no special licensing requirements to fly the Skycaster.

Waugh Electronics
Box 17184 Metro Airport
Nashville, Tennessee 37217

EXTERIOR

ALUMIGRIP

Alumigrip AA-92 dries to a high-gloss finish, helping to reduce maintenance costs. Dirt, engine exhaust soot, and bug stains can be washed off the hard finish. It can be applied to aircraft easily using standard techniques and equipment.

Alumigrip
Grow Chemical Corporation
831 South 21st Street
St. Louis, Missouri 63103

Tandy Leather Company ramskin seat cover

Pro-Tec-Prop Aircraft Covers

FAMOUS PRO-TEC-PROP AIRCRAFT COVERS

Pro-Tec-Prop manufactures two types of aircraft covers. An outside snap-on cover comes complete with installation instructions and kit. Made of a white weather resistant vinyl with a soft polyester pile backing, the outside cover increases Plexiglas life by three times. Windows also remain cleaner. Also available is a sunscreen that attaches to the inside of the aircraft window frame area with Velcro tabs. The sunscreen is made of a single-ply aluminized face reflective material and is fiberglass-reinforced for tear resistance.

Both types of cover prevent sun bleaching of upholstery and fading of instrument faces, and both extend the life of radios and gyro bearings by lowering the interior temperature build-up.

Other covers are available for the windshield alone, for the rear window, for pitot tubes, props, wheels, and canopy.

Famous Pro-Tec-Prop Aircraft Covers
P.O. Box 1551
Big Bear Lake, California 92315

GREAT LAKES AERO PRODUCTS

Great Lakes sells acrylic windshields that can be installed in your plane using a saw with a blade at least 24 teeth per inch and a high speed. Use a band saw, not a reciprocating saw such as a saber saw.

Great Lakes Aero Products
915 Kearsley Park Boulevard
Flint, Michigan 48503

STITS AIRCRAFT COATINGS

Stits has developed a complete system of new modern materials for fabric-covered aircraft. Stits uses a proprietary nontauting, nonburning system and is FAA approved.

Stits Aircraft Coatings
43rd and Fort Drive
P.O. Box 3084
Riverside, California 92519

Stits Aircraft Coating-Herothane Finish

U.S. INDUSTRIAL TOOL & SUPPLY COMPANY

U.S. Industrial Tool & Supply Company offers hand and power tools. Examples follow:

The Air Rivet Gun is used in all types of aircraft work to rivet both aluminum and steel rivets.

The Flex Shaft Angle Drill Attachment can be chucked into an air or electric tool, thereby allowing the operator to drill holes in difficult-to-reach spots.

Bucking Bars are used to back up the rivets as they are upset by the riveting hammer.

Metal Shrinker/Stretchers (nonpowered) are designed for use with light, average metals—up to and including 16 gauge.

Strap Duplicators allow operators to drill duplicate holes in overlapping sheets with a lower joggled strap that eliminates wedging of materials.

The Angle Drill Attachment Shaft is chucked into the drill and can be used for light drilling in tight spots.

The Portable Sheet Metal Punch is for limited punching power—1.2 tons or less.

Cleco Fastener Installation Tool is designed for use with all plunger-type fasteners.

U.S. Industrial Tool & Supply Company
13541 Auburn
Detroit, Michigan 48223

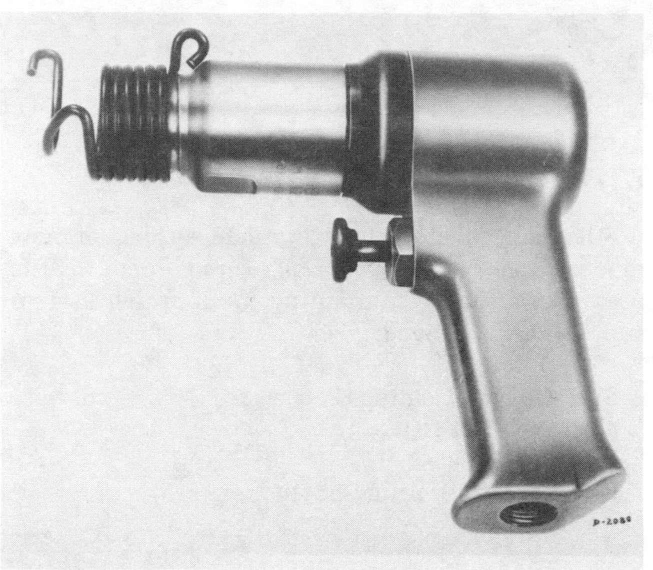

Air Rivet Gun

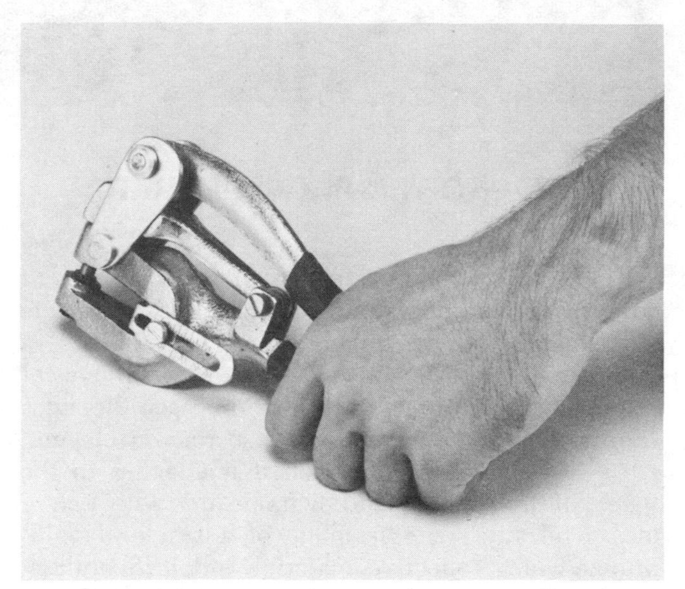
Portable Sheet Metal Punch

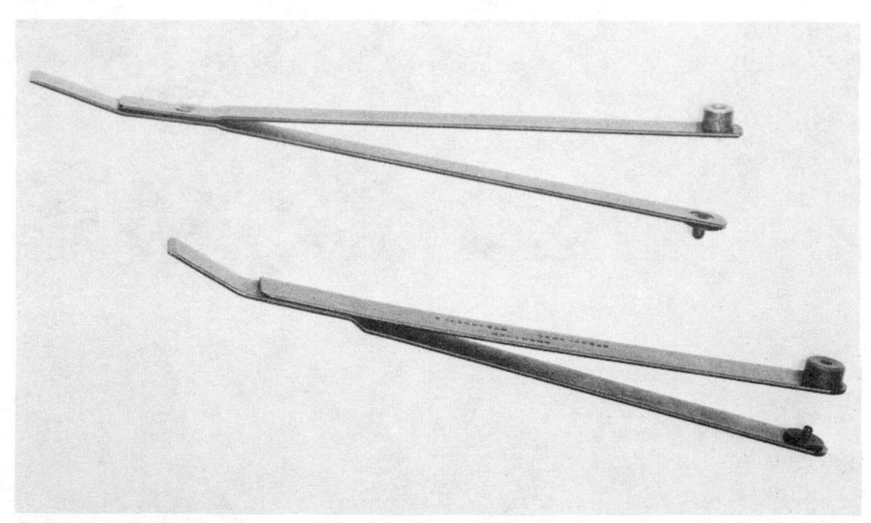
Strap Duplicators

4 / Avionics

The chapter on avionics is divided into three sections: flight instruments, radio and radar equipment, and emergency and safety equipment. But first, glossary of important avionics equipment terminology.

AVIONICS EQUIPMENT GLOSSARY

Airspeed Indicator: Instrument that registers an aircraft's speed through the air. The Mach Number states a jet's airspeed in terms of the speed of sound. The speed of sound equals 760 mph at sea level, and the airspeed equal to the speed of sound is Mach 1.0. Speeds greater than Mach 1.0 are known as supersonic speeds.

Altimeter: Aneroid barometer measuring the relative altitude of an airplane, above sea level, by measuring atmospheric pressure.

Artificial Horizon: Instrument in which a horizontal bar of a T remains fixed on the horizon despite the location of the airplane.

Automatic Direction Finder (ADF): Instrument used to receive transcribed weather broadcasts from Nondirectional Beacons (NDBs) and also used for enroute and local navigation. The ADF serves as a backup to VHF navigation equipment.

Autopilot: Automatic flight control system that keeps an aircraft on level flight or on a set course.

COMS: Radios with multiple frequencies, used by pilots to communicate with Air Traffic Control (ATC) for permission to taxi, take off and land, contact flight service stations and control towers.

Distance Measuring Equipment (DME): Instrument consisting of a distance indicator and UHF transmitter and receiver that provides the nautical mile distance between an aircraft and a VOR station. If a pilot flies a plane above 24,000 feet, it must be equipped with DME.

Flight Control System: Instrument that provides flight data and greater capability than an autopilot. It includes a Flight Director Indicator and course deviation bar and glide slope pointer of Horizontal Situation Indicator.

Marker Beacon: Marks the position of an airplane as it progresses along an airway, by providing visual and aural indications of the 75 mHz signals as each marker is passed to help pilot fix aircraft's position.

NAVS: Help pilots utilize VOR radio stations, which give access to nautical miles and give localizer capability so that a pilot can line up the airplane with the runway for an ILS approach.

Rate-of-Climb Indicator: Instrument that measures the rate of change of the altitude.

RNAV: System utilizing on-board electronic computer to help pilot fly the shortest distance between two points rather than from VOR to VOR station.

Transponder: Airborne radar beacon transmitter-receiver that automatically receives signals from an interrogator and replies with a pulse or pulse group (code), helping those on the ground keep track of a plane's location. For a pilot to fly above 12,500 feet, plane must be equipped with transponder.

Turn-and-Bank Indicator: Instrument that consists of a turn indicator and a bank indicator. The turn indicator measures the direction and rate at which an aircraft turns. The bank indicator shows the relationship between the angle of bank and the rate of turn.

VASI System: For use when a pilot has visibility. Many runways are equipped with a Visual Approach Slope Indicator (VASI) system. It consists of lights that provide a pilot with a visual flight path to the approach end of a runway.

FLIGHT INSTRUMENTS

AEROSONIC

Aerosonic manufactures flight and engine instruments for the aerospace industry. Items include airspeed indicators, altimeters, encoding altimeters, altitude alerters, fuel flow systems, gyros, transducers, manifold pressure gauges, rate of climb instruments, suction gauges, and test equipment.

Aerosonic
3312 Wiley Post Road
Carrollton, Texas 75006

ARTAIS WEATHER-CHECK

The Artais Weather-Check provides weather information, certified by the FAA for utilization under IFR. This enables an instrument approach to an airport 24 hours a day, with lower IFR minimums by elimination of any penalties for use of remote altimeter settings. The wind direction and speed information includes a report on gusting conditions and variable winds.

The three main features of the system include: (1) a computer that processes the digital signals from the instrument unit and runs diagnostic self-checks; (2) an instrument system unit that receives analog signals from sensors and conditions them for processing by the computer; (3) sensors, including two quartz pressure transducers, temperature and dewpoint sensors, wind speed and direction sensors.

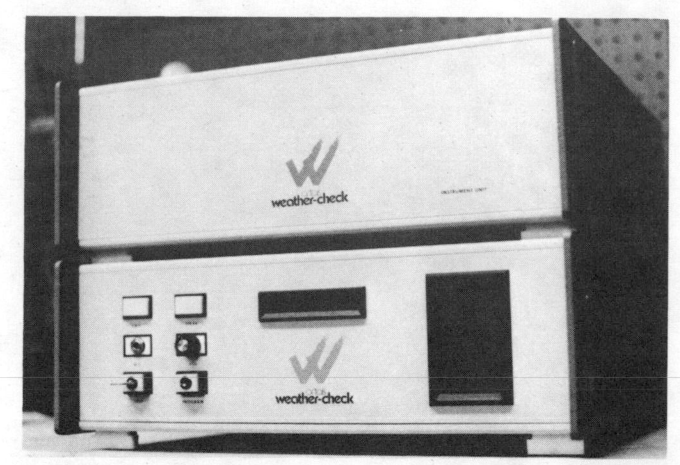

Artais Weather-Check

The following items are reported by the system: airport name, time, wind direction, wind speed, temperature in degrees Fahrenheit, dewpoint, altimeter setting, density altitude, notice to airmen.

Artais
4660 Kenny Road
Columbus, Ohio 43220

ASTRONAUTICS

Astronautics/Intelligent Displays solve complex system problems. The head-up displays, electronic countermeasures (ECM) displays, navigation displays, and aircraft data displays perform computations, change symbol formats, and manipulate graphic and alphanumeric symbols. The computer in the Intelligent Displays also solves many non-display problems: easy input-output access, up to 32K memory, and computing capability.

The autopilots from Astronautics range from a simple wing-stabilizer system to a sophisticated three-axis system. The P1 system is an all-electric wing-leveler system that provides roll and yaw sensing, stabilization functions, heading hold, and manually ordered standard-rate turns. The P2 system Pathfinder is a two-axis system that provides roll and yaw stabilization, heading hold, manually ordered standard rate turns, and VOR/localizer tracking by means of aileron servo control. The P2A system is also a two-axis system that provides unlimited capture-angle VOR/localizer, VOR/lo-

Horizontal Situation Indicators

calizer tracking with automatic crosswind correction and manually ordered standard-rate turns, all by means of aileron servo control. This unit can couple two NAV receivers with independent gain control. The P3 system provides in addition to the other systems' capabilities full-time pitch stabilization plus altitude hold via elevator control. The P3A system provides all previously mentioned capabilities, plus tracking with automatic crosswind correction and altitude hold. The P3B system is a three-axis system providing full-time roll, yaw, and pitch stabilization; heading selection and control; VOR/localizer tracking with automatic crosswind correction; manually ordered standard-rate turns; pitch stabilization in all altitudes; automatic and manual glide slope intercept; manually ordered electric pitch trim.

Some features included in Astronautics' flight director are: VOR mode with automatic capture of VOR radials from any selected heading and automatic crosswind corrections; altitude hold mode with automatic release at glide slope capture; programmed ILS approach mode; and localizer back course steering capability.

Features of the Integrated Displays are: present aircraft position displayed on multicolored moving map; large map storage capacity; television and radar display; complete digital interface or analog; map-referenced electronically generated symbology; storage of 100-page flight handbook information; performance of all radio navigation display functions; area navigation; removable cassettes; map annotation capability.

Astronautics Corporation of America
907 South First Street
Milwaukee, Wisconsin 53204

L1011 Cockpit Electronic Chart System

BONZER

Bonzer produces radar altimeters:

The Mini-Mark is a compact, no-frills altimeter that sells for less than $1,000. The single-antenna system is installed in the aircraft's fuselage or wing. The indicator fits a two-inch panel hole. Direct-reading linear scale provides dependable data from 80 to 1,000 feet. An optional DH system can be set to any altitude in increments of 100 feet.

The Bonzer Mark 10X provides dependable altitude data from 40 feet to 2,500 feet above ground level. The three-inch indicator features a high-visibility yellow needle that travels across white numerals for easy reading during instrument scan. A red push-to-test button permits checking the system's operational readiness in flight or on the ground. The DH bug can be set anywhere on the scale, providing altitude alerting in a range from 40 feet to 2,500 feet. The Mark 10X radar altimeter system including T/R unit, indicator and antenna weighs 4½ pounds.

The Impact radar altimeter system can be used by existing general aviation aircraft, although it is designed to interface with advanced digital flight management systems of future aircraft. A microstrip antenna and impact diode permit the packaging of the transmitter, receiver, and antenna into one compact unit that measures 5¼-by-4-by 4¼ inches. Digital circuits are quartz-crystal-controlled for precision, a positive signal-loss indication identifies only real signals, a three-inch panel indicator features a high-visibility yellow needle that travels across the white numerals for easy interpretation during instrument scan. The system gives dependable altitude data from 40 feet to 2,500 feet above ground level. A push-to-test button permits checking the system's operational readiness in flight or on the ground. Complete DH alerting program includes a two-second aural alert with adjustable audio level, plus an amber-lamp visual alert with adjustable light intensity.

Bonzer
90th and Cody
Overland Park, Kansas 66214

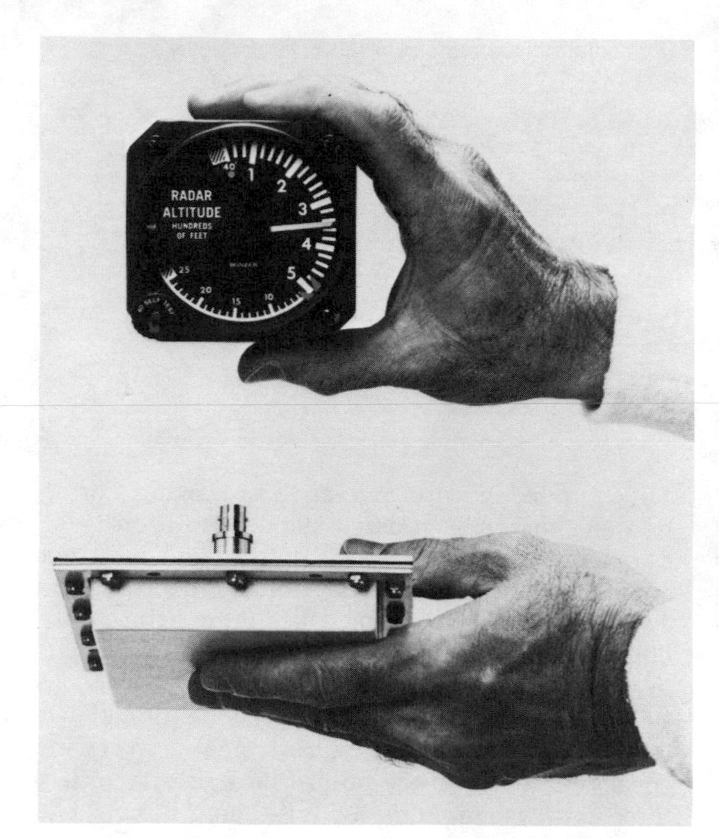

Impact Radar Altimeter

BRITTAIN INDUSTRIES, INC.

Brittain Industries manufactures automatic stabilizing and navigation control systems for single-engine and light-twin aircraft:

The Level-Matic Stabilization system provides constant two-axis control of the aircraft in all weather conditions. Straight and level flight is assured. The Turn Coordinator serves as both a flight instrument and as the gyro element, sensing and commanding attitude corrections via a pneumatic servo system. This instrument will operate from either electric or vacuum power sources.

AccuTrack acts as a stabilization system to reduce cockpit work load. Navigation tracking maintains needle-centered conditions even in extreme cross wind.

Brittain Industries Inc. Level-Matic

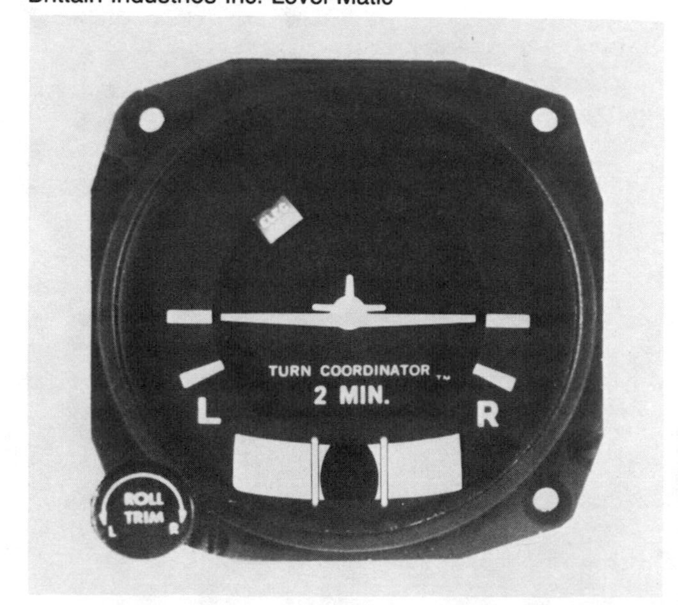

Brittain Industries Inc. AccuFlite II Control

The AccuFlite Control system allows a pilot to dial any heading, cross any radial, fly any direction. It is a lateral stability augmentation system with a heading preselect, and allows a pilot to compensate for asymmetrical aircraft loading plus heading preselect utilizing a vertical card, air-driven directional gyro with a rotatable azimuth bug.

The B-5 Flight Control system is an all-encompassing adaptable flight control system offering: electro-pneumatic operation, automatic cross wind compensation, manual turn capabilities, single-rate gyro, three-axis stabilization, magnetic heading select and hold, radio coupling, separate capture, track and approach modes, complete independence from artificial horizon and directional gyros, incomparable light weight. Components of the B-5 system include gyro sense element, controller/amplifier, power supply, magnetic heading sensor, dynertial pitch altitude control, and pneumatic servos.

The Nav-Flite II offers two-axis stabilization, turn coordinator, navigation coupling, magnetic heading, automatic crosswind compensation, programmed standard rate turns, electro-pneumatic operation.

Brittain Industries, Inc.
Hangar 12
Tulsa International Airport
P.O. Box 51370
Tulsa, Oklahoma 74151

CENTURY INSTRUMENT CORPORATION

Century Instrument Corporation manufactures accelerometers, airspeed indicators, altimeters, carburetor temperature gauges, clocks, compasses, exhaust gas temperature instruments, gyros, tachometers, and many more items.

Century Instrument Corporation
4440 Southeast Boulevard
Wichita, Kansas 67210

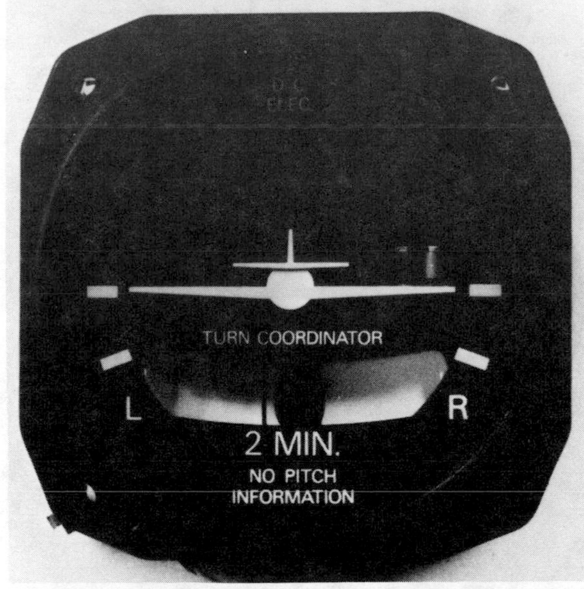

Century Instrument Corporation Flight Instruments

DAVTRON, INC.

Davtron products include the following:

Digital VOR Indicator displays VOR radial information reading in degrees "to" or "from" the station. The pilot can select Nav. 1 or Nav. 2 receiver and a "to" or "from" bearing. The bearing is automatically updated.

Digital Outside Air Temperature automatically dims, has a miniature remote probe, incandescent display, and auto polarity.

Digital ADF Indicator provides an inexpensive way to upgrade an ADF to digital operation. Model 701B will convert any nondigital ADF to digital operation and provide scan capability of dial tuning and true accuracy of a digital. The indicator comes in two models: small panel mount unit (701B-1) and 2¼-inch clock mount unit (701B-2). Front panels of both models are interchangeable.

Digital Count-Down Timer permits quick dial-in of approach time up to nine minutes and 59 seconds, retains memory of approach time start, begins automatic countdown when outer marker (or other external signal) is reached, and when count reaches zero, its automatic visual alarm starts flashing with output for an audible alarm. The unit continues counting but automatically changes to an up-count.

The Digital Voltmeter monitors the aircraft's voltage precisely. Pilot can more accurately know the amount of capacity remaining in the battery when there is a low voltage condition. Also of importance is knowledge of the voltage at which the battery is being charged. Overcharging can lead to excessive water loss and battery failure.

Davtron Digital Count-Down Timer

Davtron Digital Voltmeter

Model 202 Intercom provides noise-free, hands-off communications. Both pilot and co-pilot headsets are completely quiet until the microphone is spoken into. Each microphone is on only during the time it is spoken into. This model reduces noise by allowing each microphone to come on separately. Pilot and co-pilot have a "push to transmit" switch with priority transmit. The first switch activated has transmit capabilities. The one transmitting loses intercom side-tone, but receives side-tone from the radio transmitter verifying transmission. The second switch is completely disabled during the transmission. The one that is not transmitting hears the one transmitting at normal intercom volume. The model also has a squelch circuit and audio switching circuit that eliminate transient noises. Quality headsets should be used in conjunction with this intercom.

There is also a model MP-202, a portable intercom that plugs into the microphone and headphone jacks in an aircraft. A power receptacle plugs into the cigarette lighter. The intercom is prewired and ready to use. This model is identical to the standard nonportable 202 in operation. The portable model provides pilot and co-pilot microphone and headphone jacks on the front panel. The connector of the portable model is wired with a phone input cable, microphone cable and power lead that plugs directly into most aircraft.

Digital Pressure Altitude is a digital readout for the blind encoding altimeter, which reads altitudes from minus 1,000 feet to plus 30,700 feet in one-hundred-foot increments. It displays the altitude the transponder reports and becomes an emergency altimeter back-up system.

Digital Clock Flight Time Recorder Elapsed Time Meter displays three functions: clock, flight time and elapse time. Options include a rechargable battery and choice of converting flight time position to GMT for a total of local time GMT and elapsed time.

Davtron, Inc.
427 Hillcrest Way
Redwood City, California 94062

EDO-AIRE

Edo-Aire produces autopilots and flight directors, gyros, panel instruments and engine accessories, navigation and communications equipment for general aviation owners. In addition, many Edo-Aire products are sold directly to aircraft manufacturers for installation as original equipment on production aircraft. This equipment can be ordered through any of the Edo-Aire dealers located in the U.S., Canada, and overseas.

Some of the products include:

Century IV Series Flight Director/Autopilots include automatic pitch trim and radio and glide slope coupling in the basic Century IV autopilot. System options include yaw damping, choice of three different heading systems, and various heading/steering displays.

Century III Autopilot is a three-axis system featuring precision heading and altitude hold and is available with automatic trim, yaw damping, omni/localizer coupler, and glide slope coupler.

Century IIB and Century I are also precision autopilot systems for single-engine and light twin aircraft. System options of the Century IIB include omni/localizer radio coupler, standard or slaved DG heading systems, and yaw damping.

360 Series Navigation Situation Displays are interchangeable in their use with Century II, III, IV, Autocontrol III, and Altimatic III series autopilots without the need for interfacing adapters. All models are TSO'd and radio signals are to ARINC standards.

EDO's 660 series NAV/COM reflects high performance with the auto omni as an option. If pilot's flight needs do not require TSO'd equipment, Edo-Aire offers the companion 550 series NAV/COM. The RT-661 and RT-551 COM transceivers both deliver six watts nominal power and offer 360 channel or 720 (optional) channel capability. NAV receivers have 200 channels in 50 kHz spacing.

EDO dual set combinations of its TSO'd 660 series and 550 series offer the aircraft owner the wide choice of equipment configurations to suit individual needs and flight requirements.

Portable/mobile transceivers include the PRT-551 portable COM transceiver and MRT-551 mobile mount. The transceiver can serve as a removable radio for aircraft ferrying, field communications, or

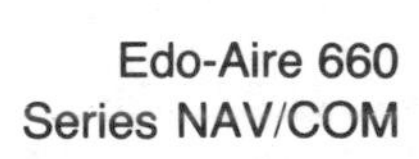

Edo-Aire 660 Series NAV/COM

Garrett Airsearch Control Display Unit

as a handy base radio. It operates from either a self-contained battery power source or from an external 12-volt source. Transmitter power is 6 watts nominal. Voice communications can be carried over 360 (VHF) channels.

The Mobile Mount offers a special mount for easy underdash installation in vehicles. Mount unit contains a speaker, quick-disconnect COM RF connector, built-in microphone jack, and a microphone storage bracket. Antennas are offered in a variety of configurations to meet vehicle requirements.

Automatic direction finders assure the pilot of receiving all radio beacons and compass locators throughout the world. Timer read-out is in seconds and minutes.

Edo-Aire offers three TSO'd ATC transponders. The new RT-887 has two pairs of concentric switches to set any of 4096 codes, bright incandescent numeral display, and decoder/encoder for long-term stability. The RT-667 and RT-777 both have fixed-tuned, super heterodyne receiver, signal decoder, reply pulse encoder, modulator, and class 1A transmitter.

DME equipment consists of a dual display cockpit indicator and a remote-mount pulse-transceiver. The indicator presents a digital read-out for distance to station (nautical miles) from NAV one or NAV two receivers.

The M-551 microphones cancel out noise, providing good voice communications. It incorporates a transistorized amplifier. Antennas feature good electrical performance, full environmental capability, and low inflight drag.

Edo-Aire encoding altimeters fulfill automatic altitude reporting. Altitude display is digital via a drum-counter and single-sweep pointer for quick altitude reference. A warning flag indicates power interruption.

The dry air pumps deliver clean dry air (pressure or vacuum) for operation of air-driven aircraft instruments. All pumps feature bi-directional drive and design improvements for exceptional reliability and long service life. Model 1U328 is specifically made for applications requiring high air pressures and flows.

Edo-Aire Model TCE 71-A area navigation system offers 20 waypoint computer memory, vertical navigation capability, automatic tuning, information monitoring, and manual data insertion. The system consists of three units: control display, automatic data entry, and remote-mount NAV computer. VOR/DME navigation systems provide the essential electronic inputs. Pilots can program the system directly by means of the control display units or insert from one to 20 waypoints by means of the card-reader automatic data entry unit. The model also incorporates features such as slant-range correction, self-check data monitoring, and flight director/autopilot guidance.

Edo-Aire also produces loran and chronometer systems.

Edo-Aire
Division of Edo Corporation
216 Passaic Avenue
Fairfield, New Jersey 07006

GARRETT AiRESEARCH

The Garrett AirNAV Series 300 allows essential navigation data to be displayed on the weather radar screen and has a CDU-control display unit with a redesigned face panel for easier programming operation.

Garrett
AiResearch Manufacturing Company
2525 West 190th Street
Torrance, California 90509

GARRETT CORPORATION

Garrett's general aviation items and services range from turboprop and turbofan aircraft engines and area navigation and environmental control systems to design and modification of customized interiors, down to valves, controls, and other small parts.

Some of the specific items produced include the following: Engine equipment includes control systems, performance indicators, speed sensors, bleed airflow sensing and control systems, turbochargers and starters. Auxiliary power units give business aircraft complete self-sufficiency. Avionics equipment includes flight navigation management systems, passenger information display, true airspeed computers. Environmental control systems include air-conditioning units, cabin pressurization systems, and temperature control systems. Survival equipment items are evacuation slides, life rafts, survival beacons for overwater operations. Test equipment produced by Garrett: turbocharger control component testers, engine over-speed circuit testers, pneumatic function controllers, cabin pressure test stands. In addition, Garrett has aviation service centers with line service with short turnaround time. Garrett also has FAA-approved repair stations.

Garrett Corporation
9851 Sepulveda Boulevard
P.O. Box 92248
Los Angeles, California 90009

GENERAL AVIATION TRADING COMPANY

The GATCO-1 Drift/Groundspeed calculator incorporates colored graphics and visual elements to allow a pilot to quickly assess the relationship of aircraft and wind. Simple multipliers allow quantitative evaluation of drift and groundspeed. The reverse side of the calculator includes an easy to use NM-SM-KM conversion scale as well as a Time-Distance-Speed chart.

General Aviation Trading Company
P.O. Box 385
Walnut Creek, California 94596

Instrument and Flight Research Inc., Slaved Gyro

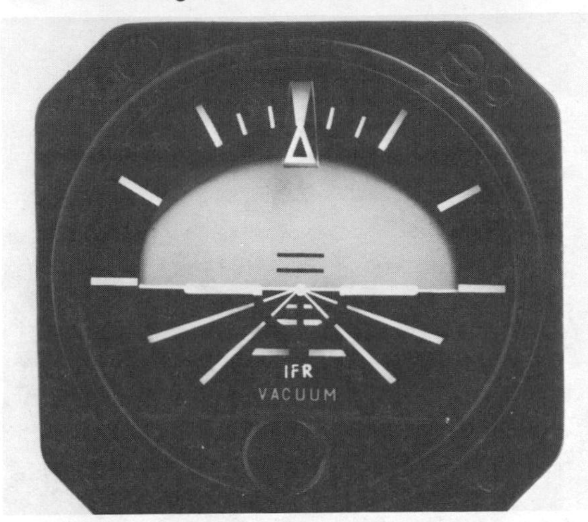

Instrument and Flight Research Inc., IFR-E41 Encoding Altimeter

INSTRUMENTS & FLIGHT RESEARCH, INC.

Instruments include slaved gyro systems and an encoding altimeter. The slaved gyro systems provide drift-free headings for precision cross-country navigation and instrument approaches. The systems come complete with indicator, flux valve detector, cable, slaving meter, and breaker switch.

The encoding altimeter is engineered for IFR flying, needs no panel modification, has an optical sensor, has three-pointer configuration.

Instruments & Flight Research, Inc.
2716 George Washington Boulevard
Wichita, Kansas 67210

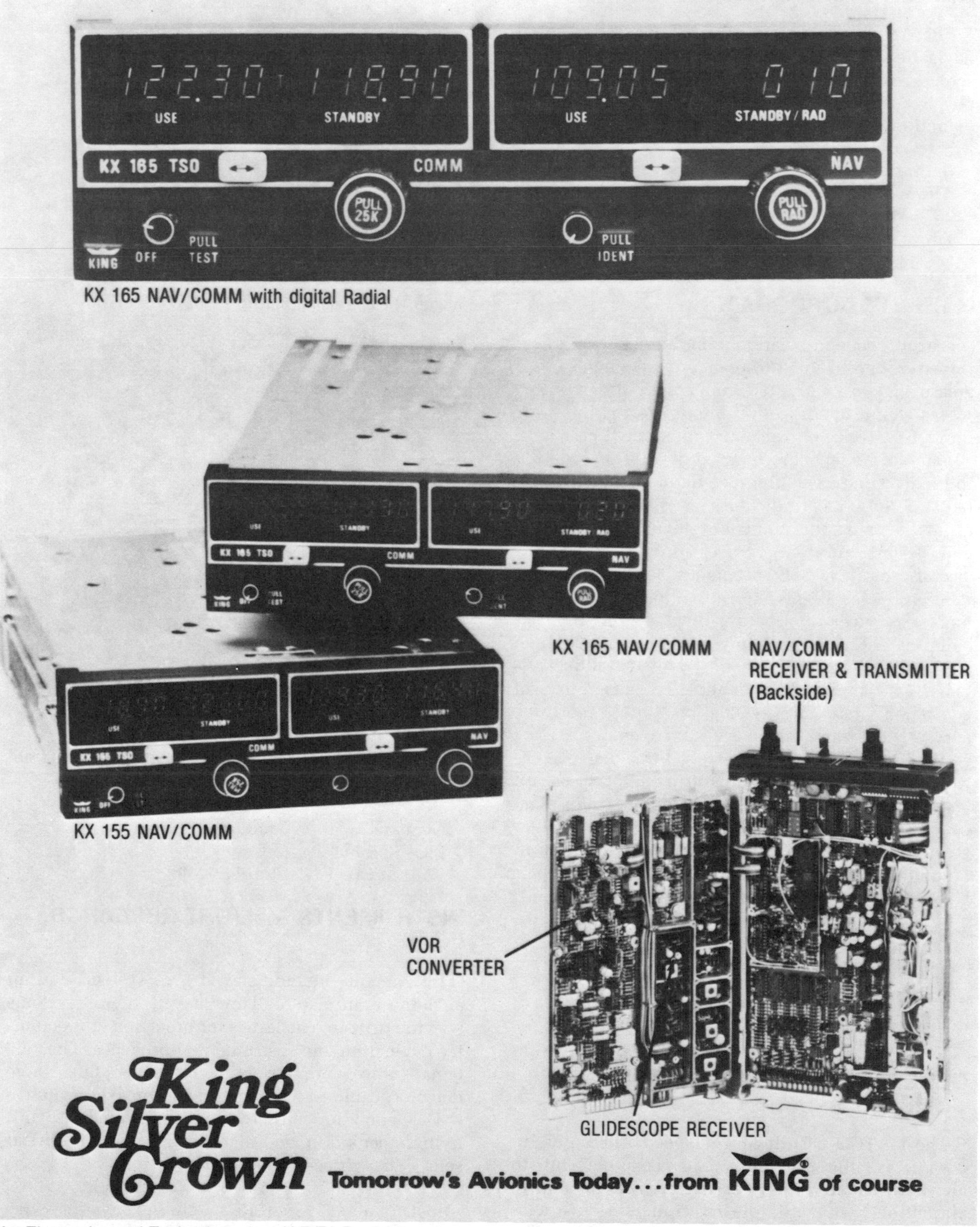

Jet Electronics and Technology, Inc. (J.E.T.) Products

JET ELECTRONICS AND TECHNOLOGY, INC. (J.E.T.)

Products from J.E.T. include aircraft control systems (automatic flight control systems, yaw dampers, mach trim and spoileron computers, nosewheel steering and fuel monitoring, torquer actuators) and gyros and flight instruments (vertical and directional gyros, rate gyros and turn and slip indicators, self-contained attitude indicators and flight directors). Examples include:

Two-inch gyro horizon, Model A1-804, serves as a standby reference indicator for high performance aircraft.

Three-inch gyro horizon, Model A1-904, is an attitude reference indicator.

Jet Electronics and Technology, Inc.
5353 52nd Street
Grand Rapids, Michigan 49508

KOLLSMAN INSTRUMENT COMPANY

Kollsman, a division of Sun Chemical, produces the Alti-Coder II for altitude reporting. The altimeter senses atmospheric pressure, displays altitude with high accuracy, and generates an altitude reporting signal. It functions as a pneumatic altimeter in the event of a power failure. A knob on the front of the altimeter provides the means for setting barometric pressure.

Kollsman Instrument Company
Division of Sun Chemical Corporation
Daniel Webster Highway South
Merrimack, New Hampshire 03054

Kollsman Instrument Company Alti-Coder II

MET-CO-AIRE

The main products of Met-Co-Aire are wing tips and wing tip fuel tanks, both constructed from reinforced fiberglass. The tips give an increase in cruise speed, an increase in rate of climb, reduce stall, decrease the distance required for take-off, and give a new, sleeker profile to the aircraft. The tips are available for Piper Aztecs and Apaches, Cherokees, Comanches, single-engine Cessnas, and Bonanza 35s.

The fuel tanks are a combination of the tip and a long-range fuel tank, giving the pilots the benefits of the tips and also having extra fuel on board in the tanks.

Installation is easy, requiring approximately two hours. Tips are FAA-approved (Supplemental Type Certificated).

Met-Co-Aire
P.O. Box 2216
Fullerton, California 92633

NARCO AVIONICS

Narco manufactures a variety of avionics equipment, including transponders, altitude encoders, audio control panels, NAVs and COMs, marker beacons, glide slope receivers, direction finder systems, and distance measuring equipment (DMEs).

CP 135 and CP 136

Both units employ 100 percent solid-state technology and components. Pushbutton controls with indicators give clear annunciation of the selected audio input. Interlocking microphone/audio controls allow single-action switching from one Com

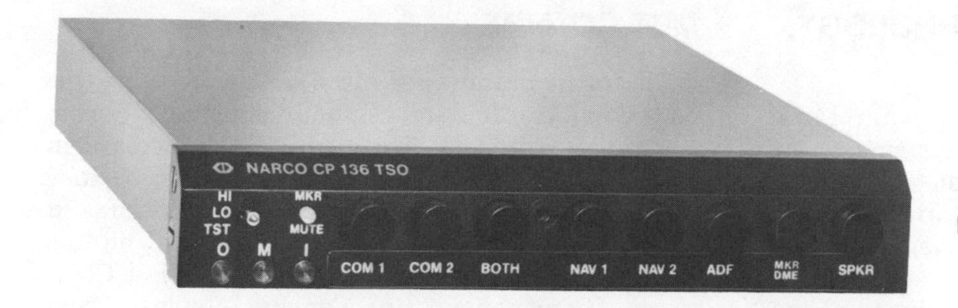

Narco Centerline CP 136

transceiver to another, while an audio selector lets the pilot simultaneously monitor both Com receivers. A single pushbutton turns cabin speaker audio on or off; headphones audio remains on at all times.

The CP 136T, a special audio panel, is designed for use in aircraft equipped with both VHF and HF transceivers, or where independent control of speaker and headphones audio inputs is desired.

NAV-122, Centerline NAV-124, and Centerline NAV-121

The NAV-122 is a complete system, independent of other avionics. It has its own frequency tuning controls and readout, rotable OBS card, on/off IDENT volume control, VOR/LOC course deviation indicator, and precise rectilinear glideslope needle.

The NAV-124 includes a built-in VOR/ILS indicator.

The NAV-121 was designed for those pilots who do not want or need glideslope and marker beacon capabilities in their aircraft. It can be used as a primary system in VFR-equipped aircraft or as a back-up to a NAV-122 in full IFR equipped panels.

COM-120

The COM-120 has a full six-digit, easy-tune readout, transmits a monitor light, and has 10–13 watts transmitter output. It is full/ TSO'd with 720 channels.

Narco Centerline NAV 122 and NAV 121

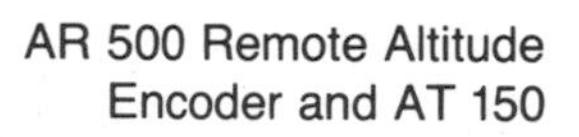
AR 500 Remote Altitude
Encoder and AT 150

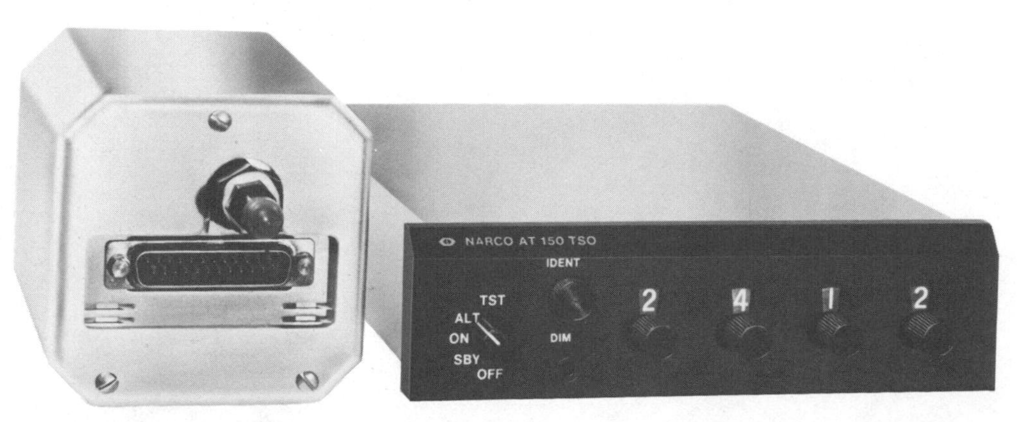

MKR-101

The MKR-101 is the solution to those applications that call for a marker beacon receiver separate from a primary NAV unit. In its remote-mounted version, the MKR-101 will drive the marker lights located in the Centerline CP-135 and CP-136 audio panels, as well as marker lights in many NAV indicators, including the Centerline ID-124. The MKR-101 also offers a high/low sense and "on-off" switch plus pilot-actuated audio muting.

ID-24

A TSO'd course deviation indicator, the ID-124 provides rectilinear meter movements, marker beacon lights, nonglare antireflective glass, self-test pushbutton for VOR and marker lights, front or rear panel mounting.

UGR-24 2-5

A glide scope receiver, the UGR-24 2-5 can be channeled by standard ARINC 2 out of 5 systems such as the Centerline NAV-122 and NAV-124 series, while retaining its ability to be channeled by competing NAV receivers that use other code systems.

ADF 141

The ADF 141 Direction Finder system has a full range of ADF and AM broadcast bands, 200-179-kHz. IDENT switch enhances station audio and allows easier, assured I.D. It also includes an antenna coupler for better range and clarity than combined sense/loop antennas and autopilot outputs.

DME-195 and DME-190 TSO

The compact panel of the DME-195 provides simultaneous display of distance (up to 200 miles), groundspeed (up to 400 knots), and time-to-station (up to 89 minutes). Also located in the panel display area are NAV 1/NAV 2/DME HOLD selector switch, RNAV indicator light, on/off IDENT volume control and push-to-test and display dimmer controls.

The DME-190 is also compact, gives a display of distance up to 200 nautical miles, has time-to-station up to 89 minutes, and groundspeed up to 400 knots. An optional Remote switch lets the pilot channel through an associated NAV receiver.

Information on other Centerline avionics products is available from the manufacturer.

Narco Avionics
Narco Scientific Industries
270 Commerce Drive
Fort Washington, Pennsylvania 19034

ROCKWELL INTERNATIONAL, COLLINS GENERAL AVIATION DIVISION

The Collins General Aviation Division supplies avionics systems, products, and services. The Pro Line consists of remote-mounted avionics designed for use in heavy twin-piston engine, turboprop, and business jet aircraft. The Micro Line includes panel-mounted avionics designed for use in aircraft ranging from heavy single-engine through medium twin-piston engine aircraft, including helicopters.

TDR-950 Transponder is a panel-mount transponder with the full 4096 codes and altitude reporting capability up to 62,000 feet. It is TSO'd to the highest class for unrestricted service.

ALT-50/55 Radio Altimeters are designed for the business aviation fleet. The ALT-50 reads from 2,000 feet to touchdown while the ALT-55 reads from 2,500 feet, so it is compatible with ground proximity warning systems. Both altimeters meet TSO requirements.

LRN-85 features alphanumeric inputs and outputs, permanent mass memory bank of up to 4,000 fixed locations for use as waypoints, E or H field antennas, flight planning compute mode, continuous display of all stations present. The navigation system also receives all four format and eight unique Omega frequencies, plus all VLF transmission formats—up to 80 possible carriers.

VHF-251 is a panel-mounted communications

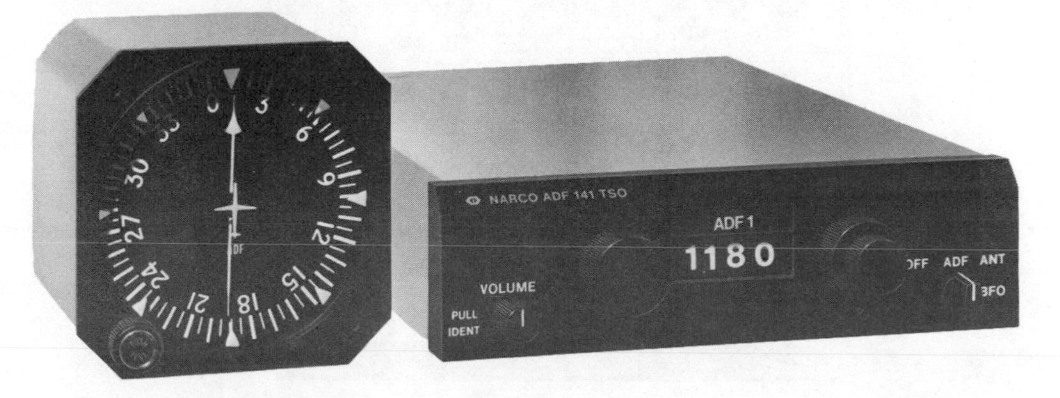

Centerline ADF 141 Indicator and Receiver

Collins Micro LIne TDR-950

ALT-50/55 Radio Altimeter

VHF-251 Communications Transceiver

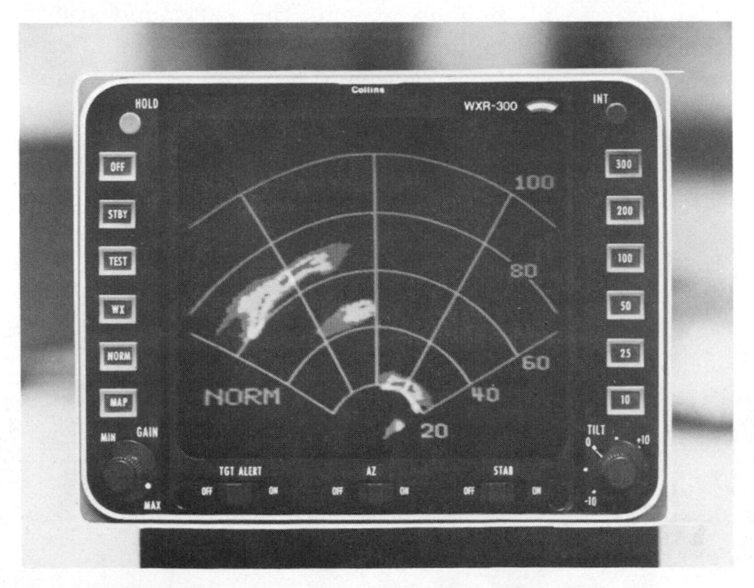

DME-40 Distance Measuring Equipment

transceiver. The 3.4-pound transceiver provides 720 communication channels in 25-kHz steps from 118,000 to 135.975 MHz.

DME-40 Distance Measuring Equipment can be mated to any one of four indicators with the 339F-12A providing dual nav capability, frequency hold, ground speed, and time-to-station. The unit, only 8.8 pounds, has indicator/distance, groundspeed and time-to-station readouts, 250-nautical mile range, all 252 DME channels, instant lock-on.

WXR-300 Color Radar has pushbutton control of both mode and range. The system automatically selects the best pulse width for the selected range. It also can contour long-range weather.

The RNS-300 allows a pilot to instantly plot an efficient course or heading around cells without trial and error. Before takeoff, the system lets a pilot map a standard instrument departure or enroute course to two waypoints on aircraft equipped with dual RNAVs. It accepts inputs from two VOR/DMEs, RNAVs, or VLF/OMEGA nav receivers, and two HSIs for course and heading control to display navigation data.

APS-841H Autopilot can function as a work reliever and can establish precise headings over long periods of time. Important features of the unit include: parallel servos as a backup in case of hydraulic failure, servo change accomplished by maintenance personnel in less than one hour, 100 percent authority so no trim actuators and no position monitoring.

EFDS-85 Electronic Flight Director System formats include Collins V bar steering and cross pointer displays on EADI model and conventional HSI or expanded scale HSI with weather radar returns on the EHSI model. The EADI display includes radio altitude, pilot selected decision height, V bar steering command, and glide slope deviation needle. The conventional HSI presentation on the EHSI includes full compass card, selected heading of 081 degrees and selected course of 043 degrees, and glide slope deviation needle. It is also possible to display two bearing pointers on the conventional HSU presentation or to switch HSI to expanded scale consisting of a 90-degree sector on the compass card and weather radar returns.

Micro Line IND-451 DME Indicator is a three ATI digital indicator, included with the Collins Micro Line DME-451. It includes an on/off switch and mode selection of NAV 1, NAV 2, HOLD, or RNAV. The HOLD function keeps the DME frequency as selected even after VOR frequency is changed. A top window always displays distance while a bottom window indicates mode status and either elapsed time, time to station, ground speed, GMT or whatever time is set in the system and ETA (time to station plus GMT setting). A stopwatch function starts when a white pushbutton is depressed.

Micro Line IND-450 DME Indicator is part of the

APS-841H Autopilot

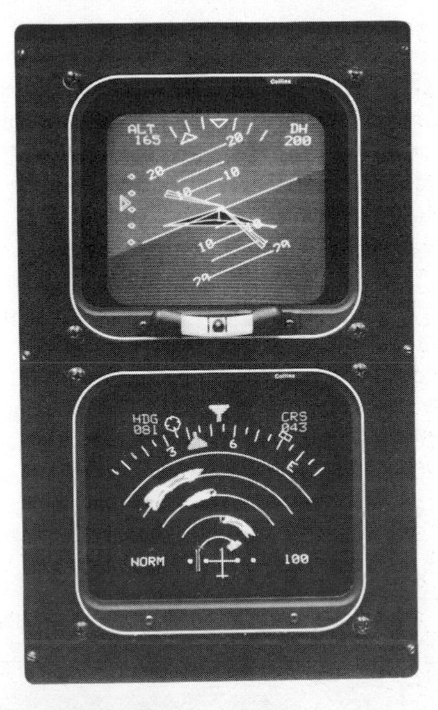

EFDS-85 Electronic Flight Director System

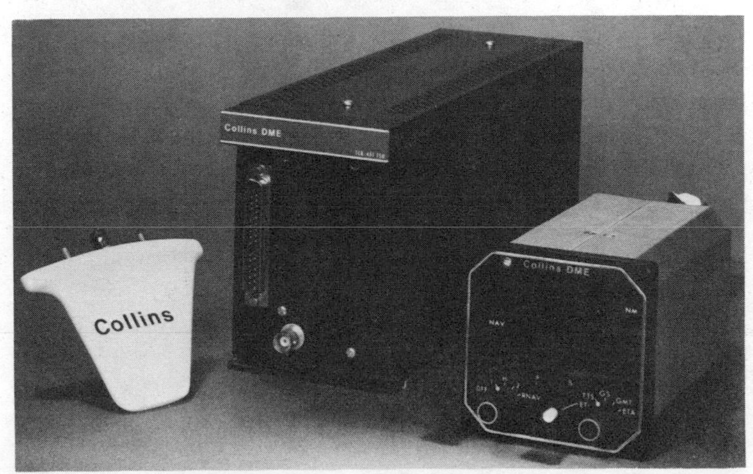

Collins Micro Line DME-451

Collins Micro Line DME-450 system, designed for applications where panel space is scarce in a single or a twin. The IND-450 features a single incandescent digital display that is shared between nautical miles, knots, and time-to-station information. An RNAV annunciator light shows if the distance displayed is to an RNAV waypoint.

Rockwell International
Collins General Aviation Division
400 Collins Road NE
Cedar Rapids, Iowa 52406

RYAN STORMSCOPE

Stormscope is a receiving system with specialized computer processing capabilities that receives electrical signals generated by storm turbulence and maps the activity for pilot viewing. The primary purpose of the system is thunderstorm avoidance. A small green dot on the CRT display is produced by each electrical discharge, indicating direction and distance relative to aircraft heading. As additional signals are received, a map of electrical activity is formed presenting the pilot with view of thunderstorm activity 360° around the aircraft for a range of more than 260 nautical miles. As electrical activity continues, more dots are displayed. Even though electrical signals come from every direction, each signal is individually processed and displayed to give the pilot an accurate representation of the thunderstorm activity.

The system consists of three interconnected electronic boxes plus an antenna. The three boxes are termed: receiver, display and processor/computer. The receiver and display units are mounted on the aircraft instrument panel. The computer/processor is mounted remotely within the aircraft. The three units are then interconnected by electrical cables. Information from the receiver is dispatched to the computer/processor. There mathematical functions are performed, arranging electrical images into a maplike fashion for display purposes. Further signals are held in a digital memory and updated as required. The image is connected to the display unit, which presents the pilot a map of convectively generated electrical activity within almost 300 NM of the aircraft's position in any direction from the aircraft.

With the Stormscope antenna, there is no aerodynamic drag. Installation can be made on any fixed-wing or rotor-wing aircraft, regardless of size, type or speed. The antenna is used on all Stormscope Weather Mapping systems.

Ryan Stormscope
4800 Evanswood Drive
Columbus, Ohio 43229

SAFE FLIGHT INSTRUMENT CORPORATION

Safe Flight specializes in Angle-of-Attack/Stall Warning equipment. The Safe Flight SC-150 system measures the stagnation point movement and translates angle-of-attack information onto a "Slow-Fast" indicator. High angles-of-attack bring needle movement toward the "Slow" diamond and into the red or stall area on the indicator. Low angles-of-attack bring the needle movement toward the "Fast" diamond. The SC-150 is also especially helpful in rough air, providing easier control and greater stability than the normally erratic airspeed indicator.

The I-500 Speed Indexer Light System provides the

Ryan Stormscope

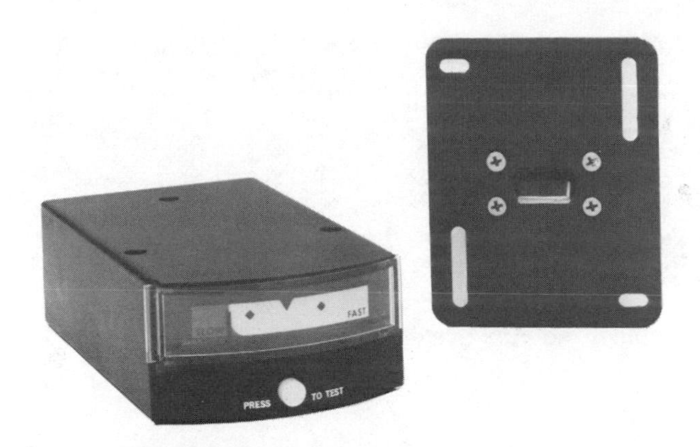

SC-150 Indicator / Lift Transducer

pilot with a direct and peripheral indication of the aircraft's speed/lift condition. The display is based on a signal supplied from a Safe Flight Computer. Excessively slow speed/lift conditions are displayed by a blinking red "S" (Slow) light.

The Scat/Autopower Speed Command Automatic Throttle System provides aircraft operators with the flight-proven features of consistently safe directed takeoffs and go-arounds, plus the benefit of automatic throttle control on landing approach.

The Fuel Performance Computer computes wing load (weight) and altitude and the optimum speed target for maximum-range or long-range cruise.

The Wind Shear Computer anticipates hazardous low-level wind shear, combines horizontal and vertical components of wind shear/downdraft, displays energy loss due to wind shear, and is designed to prevent false warnings.

Safe Flight manufactures other products for general aviation.

Safe Flight Instrument Corporation
P.O. Box 550
White Plains, New York 10602

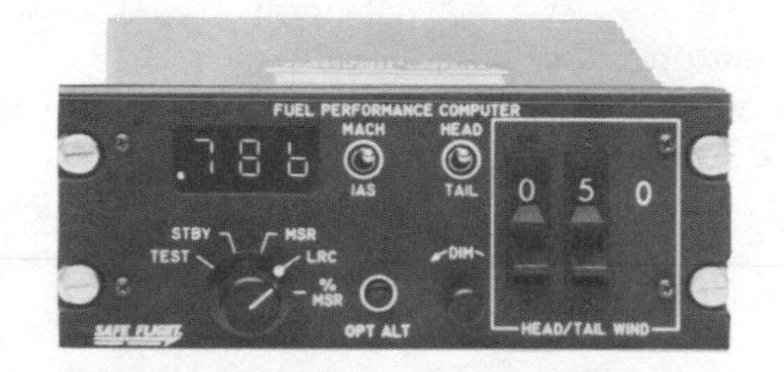

Performance Computer / Performance Controller

SPECIALIZED ELECTRONICS CORPORATION

Navtronic is an electronic computer designed for aviation that comes in a variety of models:

Navtronic 1701tr includes a built-in timer/alarm and RNAV program. It is preprogrammed to solve problems on the FAA instrument written test.

Navtronic 1701tr has a built-in timer/alarm to check "lag time" and approach time. It is useful for nonprecision approaches and fuel management.

Navtronic 1701r has a built-in RNAV program, telling the distance and course between two waypoints when both are adddressed by the same VOR.

Navtronic 1701 has all standard programs. The timer/alarm and RNAV options can be added later. The model is preprogrammed to solve all problems on the FAA instrument written test.

Specialized Electronics Corporation
9629 Irving Park Road
Schiller Park, Illinois 60176

Navtronic 1701tr

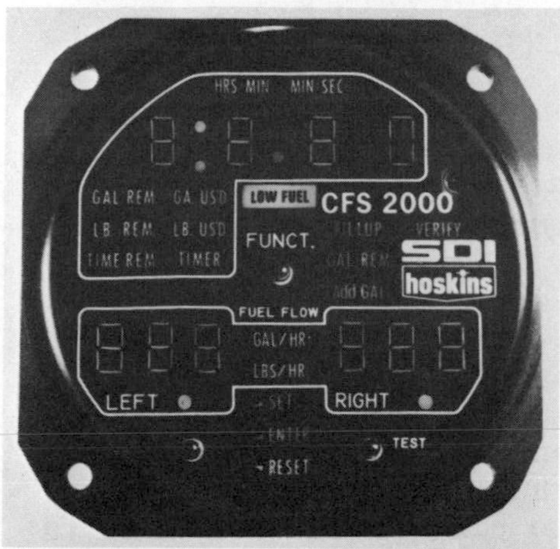

CFS-200 Flatpac Computerized Fuel System

CFS-2001 Flatpac Computerized Fuel System

SYMBOLIC DISPLAYS, INC.

The CFS panel-mounted instruments by Symbolic contain all system electronics, operating/programming controls, and digital readout display. The CFS computer counts precisely the number of pulses from the fuel flow transducer(s) and converts the count to gallons. A clock reference computes the fuel flow rate and timer read-out functions.

The products provide the pilot with the following time and fuel management capabilities: fuel flow/pounds or gallons, gallons remaining, pounds remaining, time remaining, gallons used, pounds used, clock/timer. The units are designed for use in single and twin engine aircraft.

Symbolic Displays, Inc.
1762 McGaw Avenue
Irvine, California 92714

RADIO AND RADAR EQUIPMENT

AIRCRAFT RADIO AND CONTROL

Aircraft Radio and Control (ARC) is a division of Cessna Aircraft Company, designing and manufacturing avionics equipment. There are three basic "lines" of ARC Avionics: the 300, 400, and 1000 series.

1000 Series

The 1000 COM and NAV units feature dual frequency readouts, automatic frequency storage, and three-frequency management. The 1000 NAV system provides automatic autopilot course datum to the autopilot or IFCS.

The 1000 ADF allows pilot to preselect frequencies and have them at a touch of the button. When the pilot puts up a new frequency, the frequency last used automatically goes into storage. It can be brought back with a touch of the recall button.

1000 RMI combines VOR (omni) and ADF bearing information on one indicator with magnetic heading information from an ARC slaved compass system or HSI. This allows the pilot to rapidly determine his relation to the selected VOR and ADF. Course interception procedures are simplified for ADF instrument approaches and for flying a DME arc to a VOR or ILS approach.

The 1000 Glide Slope has a sensitive 40-channel receiver and is designed to be compatible with the serial data system in the 1000 NAV.

The 800 series Transponder allows the pilot to change transponder codes with quick-select levers.

The 800 series Encoding Altimeter/Altitude Alert system eliminates erratic altitude indications commonly found with mechanical drives. The altitude alert system gives an amber light and audio signal at 1,000 feet from above or below the selected altitude, and an amber light with audio warning when the altitude deviates more than 300 feet from the selected altitude.

The 800 series DME is compatible with the serial data system of the 1000 NAV. It offers ground speed and time-to-waypoint capability when coupled with the 800 RNAV.

The 800 series RNAV is a five-waypoint system. Displays flash whenever DSPL and FLY selectors are out of phase indicating that displayed waypoint data and actual waypoint being flown are not in agreement.

1000 Audio Panel has a switching panel for a full complement of communications and navigation equipment. The master volume control allows the pilot to set the individual radio volume controls just one time.

300 Series

The 300 series NAV/COM system offers digital electronic frequency display. The 720-channel COM has a 25/50 kHz selector switch. The 200-channel NAV functions independently and provides glide slope and DME channeling. System comes with remote VOR/LOC indicator.

Automatic Radial Centering is offered as an option of standard 300 indicators. It provides three separate functions from one instrument with a three-position course selector knob.

The 300 series ADF has digital tuning for fast, accurate switching. The indicator has high-torque servo-drive for quick readings. A rotatable azimuth card is also featured.

The 300 Transponder is TSO'd to 15,000 feet and panel mounted.

The 200A Autopilot is an all-electric, single-axis aileron control system that provides lateral and heading stability. The system consists of a turn coordinator, computer amplifier, and an actuator. Push-button switches select direction hold, NAV capture, NAV track, hi-lo sense, and back course functions.

The 300A Autopilot is also a single-axis autopilot. A directional gyro with heading "bug" provides absolute heading reference and selection.

400 Series

The 400 series NAV/COM has many features of the 1000 series equipment. It features automatic frequency storage. The RT-485A holds three preselected COM and three NAV frequencies in storage for immediate recall on demand. The 720-channel COM features a green transmit light, automatic audio leveling and internal speaker amplifier. The 200-channel NAV features a three-inch rectilinear ATI indicator with automatic radial centering. The 400 NAV/COM and 400 NAV indicator provide automatic autopilot course datum when installed with slaved DG or HSI and appropriate autopilot or IFCS.

The 400 series ADF is TSO'd and has a dual-selector ADF that lets the pilot preselect two different frequencies so he can plot a fix quicker.

The 400 series RMI combines VOR (omni) and ADF bearing information on one indicator with magnetic heading information from a slaved compass system or HSI.

The 400 series Glide Slope can be used with any NAV receiver having ARANC 2 x 5 channeling characteristics, including any 300 or 400 series NAV/COM.

The 400 series Marker Beacon has a remote-mounted receiver and a panel-mounted indicator. The marker beacon for conventional twins features an optional "marker mute" button that shuts off the audible tone for 30 seconds and then automatically restores volume for passage over the next marker.

The 400 series Transponder is for high-performance aircraft operating at altitudes to 30,000 feet.

The 400 Encoding Altimeter is a servo-drive altimeter that eliminates erratic altitude indications commonly found with mechanical drives.

The 400 DME gives distance, ground speed, and time-to-station with 200 channels and 200-nautical mile range. Electronic digital readouts give the distance almost instantaneously—one-half second is the typical response time.

The 400 RNAV is a three-waypoint system that displays distance to waypoint up to 200 miles on

800 Series Encoding Altimeter/Altitude Alert System

the DME indicator. Waypoint data is entered that displays the waypoint being used or the waypoint stored as selected by the display switch. The third waypoint is preselected with the quick-select levers. The waypoint in use is selected by the Fly switch. An annunciator indicates when the waypoint displayed is not the waypoint being flown. Computed distance is displayed on the DME indicator and course deviation is presented on the NAV indicator.

Aircraft Radio and Control
Division of Cessna Aircraft Company
P.O. Box 150
Rockaway Valley Road
Boonton, New Jersey 07005

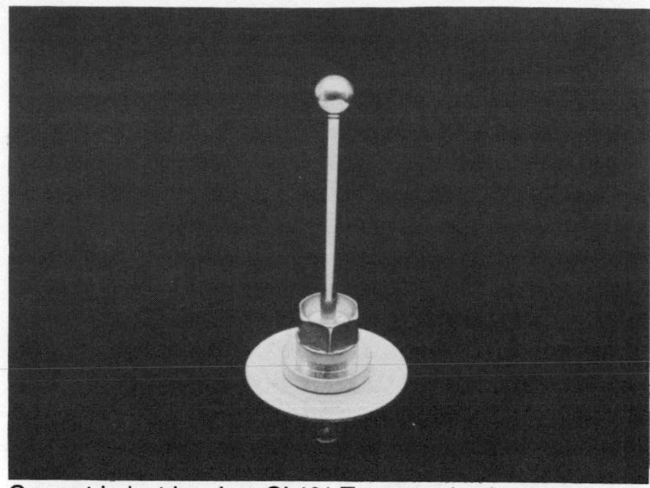
Comant Industries, Inc. CI 101 Transponder Antenna

BRELONIX, INC.

Brelonix supplies HF radios:

MODEL	FREQ.	# OF CHANNELS	PWP OUTPUT (IN WATTS)	MODES OF OPER.
SAM 100	2–14 mHz	10*	100 PEP	SSB, AME
SAM 70	2–14 mHz	5**	50 AME 40 PEP 20 AME	SSB, AME
SAM 1000	2–23 mHz	NA	150 PEP 100 AME	SSB, AME, CW (modulated without carrier)

*Simplex
**Two channels may be duplex

Brelonix, Inc
106 North 36th Street
Seatttle, Washington 98103

COMANT INDUSTRIES, INC.

Comant designs and manufactures antennas and related components for general aviation aircraft. Examples follow:

CI 100—All Metal Blade DME/Transponder Antenna

The CI 100 is an adaptation of an all metal blade designed and developed for Mach 3 military aircraft. The design is unique in that the blade and mounting base are diecast as one piece with no dielectric material in the airstream.

Freq: 960-1220 mHz; VSWR: 1.5:1 max; Height: 1½″; Weight: .25 lbs.; TSO C66a, C74b, Class 1A, DO-138; Env. Cat. BAJ XXXXXXXXX

CI 101, Transponder Antenna

The CI 101 is a top loaded stub monopole. The antenna radiator is mechanically captivated and machined from beryllium copper for impact resistance. All metallic parts are plated with bright nickel for corrosion protection.

Freq: 1030-1090 mHz; VSWR: 1.3:1 max; Height: 2½″; Weight: .1 lbs.; TSO C74c, DO-138 Env. Cat. BAJ XXXXXXXXX.

CI 102—Marker Beacon Antenna

The CI 102 is designed for use with modern, high sensitivity marker beacon receivers. The antenna is small and lightweight, featuring four hole mounting for simple installation. The antenna assembly is encased in an injection molded, glass reinforced polyester material which is impervious to the environment associated with the underside of an aircraft.

Freq: 75 mHz; VSWR: 1.5:1 max; Height: 2¼″; Weight: .6 lbs.; TSO C35d, Class 1A, DO-138; Env. Cat. AASXXXXXXXXX.

CI 109—VHF Communication Antenna

The CI 109 is a broadband communication antenna exhibiting excellent electrical characteristics. The antenna incorporates an aerodynamically shaped mounting base and radiator housing that essentially matches the styling of the communication antennas used on the 1968-72 Cessna single engine models.

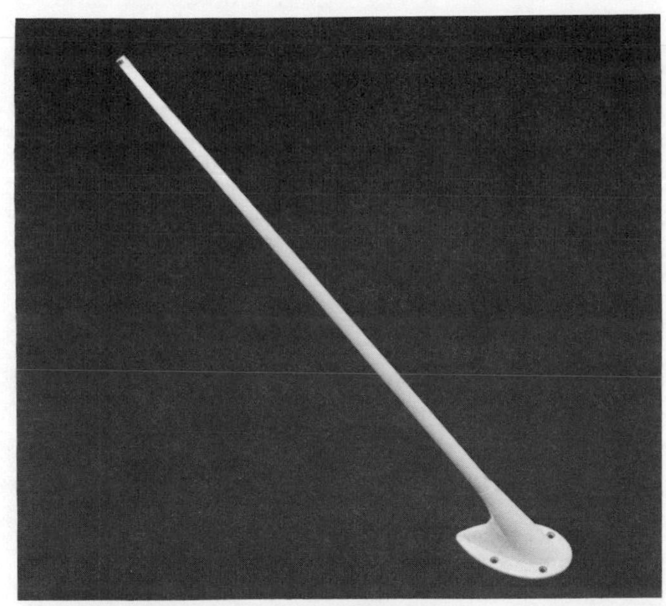

CI 109 VHF Communication Antenna

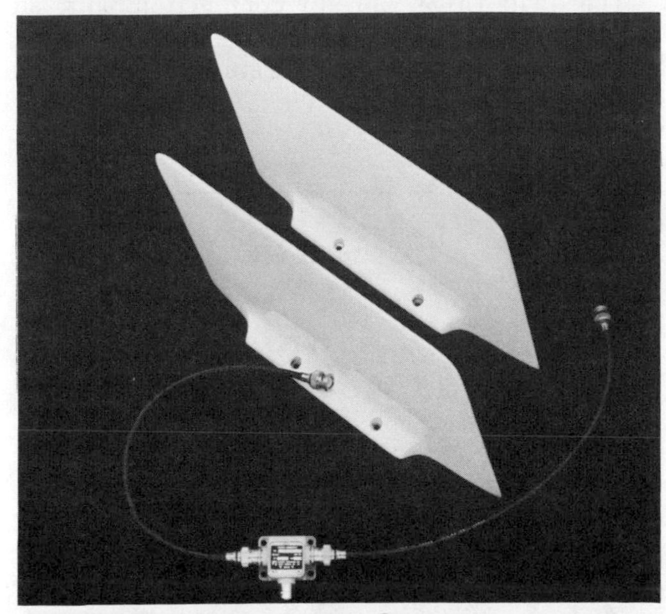

CI 120 VOR Navigation Antenna System

CI 157B V'Dipole VOR Localizer/Glide slope Antenna

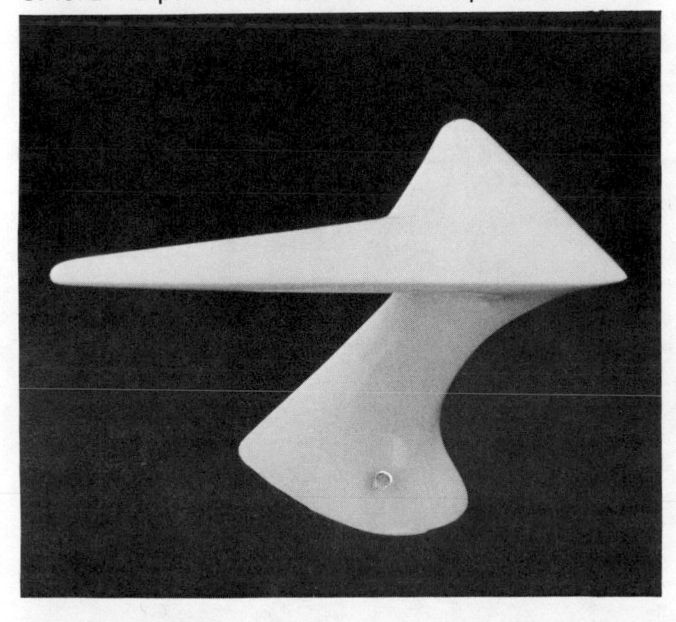

Freq: 118-136 mHz; VSWR: 2.0:1 max; Height: 17¼″; Weight: .6 lbs.; TSO C37b, Class 1, C38b, DO-138; Env. Cat. BAJ XXXXXXXXX.

CI 120—VOR Navigation Antenna System

The CI 120 is designed for optimum VOR performance when used for Area Navigation. The antenna system is qualified for use on single, twin, jet and helicopter aircraft. The CI 120/GS features glide slope capability. Dual VOR, Dual Glide Slope operation is available when used with the CI 1125 diplexer.

Freq: 108-118 mHz; VSWR: 3.0:1 max; CI 120/GS: 328-336 mHz; VSWR: 3.0:1 max; Height: 6.00″; Weight: 3 lbs.; TSO C34c, C36c, C40a, DO-160; Env. Cat. D2ADXXXDXSXXXXX.

CI 157B—"V" Dipole VOR Localizer/Glide Slope Antenna—Detachable Elements

The CI 157B is designed specifically for compatibility with Piper Aircraft Mounting. The antenna features a miniature ferrite balun integral with the mounting disc. The detachable radiating elements can be assembled after installation. Dual VOR and single glide slope operation is available with the CI 505 Diplexer.

Freq: 108-118 mHz; 328-336 mHz; VSWR: 3.0:1 max; Weight: .35 lbs.; Mounting: Vertical; Fin Stabilizer; Not qualified for use on helicopter aircraft.

CI 132—ADF Sense Antenna

The CI 132 is designed to eliminate precipitation static in the ADF system. The antenna consists of a corrosion proof tension unit, 21 feet of high tensile strength insulated antenna wire, and a simple fuselage feed through.

CI 106—Radio Telephone Antenna

The CI 106 is a stub antenna designed to withstand the environment associated with the underside of an aircraft. The antenna radiator is mechanically captivated and sealed against leakage.

Freq: 450-470 mHz; VSWR: 1.5:1 max; Height: 6¼″; Weight: .25 lbs.

Comant Industries, Inc.
3021 Airport Avenue
Santa Monica, California 90425

COMMUNICATIONS COMPANY

Communications Company manufactures remote controllers, line terminating units, multichannel receiver types, VHF AM receivers, drive units, transmitter/receivers, scanning monitor receivers, UHF AM receivers. Examples follow:

Remote Controller Type J

The controller J is designed to operate transmitting and receiving equipment remotely via a two wire and earth telephone line. Control is effected by applying 50v dc between the respective lines and earth.

The controller is housed in a small sloping fronted case intended for desk top use in conjunction with desk microphone, telephone handset or headset and boom microphone.

For use in the United Kingdom the controller may be connected to the telephone line via G.P.O. approved Park Air Line Isolator type PA1811.

Park Air 1201 series all-solid state VHF AM Receiver

Features: all solid state, single conversion, designed for high MTBF low MTTR, full metering facilities, built-in monitor loudspeaker, AC or DC operation, excellent two-signal interference protection.

The receiver is intended for rack mounting in Airport or en route stations and may be operated from standard AC supplies or 24V DC. An auto no-break changeover facility is incorporated which will cause the receiver to switch to DC supplies if the AC mains fail.

SPECIFICATIONS

Frequency coverage:	118 - 136 mHz + 10% overlap A3 Amplitude Modulation
Number of channels:	One

Drive/Units/Types 1230/1240/1210/

The 1230 Drive Unit provides a fully synthesized 720 channel drive source for the company's range of VLF communications equipment.

SPECIFICATIONS

Number of channels:	720
Frequency range:	118-136 mHz.

The 1210 Drive Unit in conjunction with an associated transmitter will provide a multichannel facility that is especially useful as emergency standby equipment.

SPECIFICATIONS

Frequency Coverage:	118-138 mHz.
Number of channels:	Up to eight.

Park Air 1201 Series VHF AM Receiver

The 1240 unit is designed to provide a single frequency high-stability VHF drive source for the company's range of transmitters. Phase-lock techniques are used to provide a spectrally pure output with a high order of stability.

SPECIFICATIONS

Number of channels:	One.
Frequency:	Anywhere within the Airmobile band 118–136 mHz.

Transmitter/Receiver Type 1300

SPECIFICATIONS

Frequency coverage:	116-138 mHz.
Service:	A3 amplitude modulation
Number of channels:	One
Power outputs:	10 watts minimum carrier (40W p.e.p.). Front panel Transmit lamp does not illuminate if power drops below 7W.

1500 Series 50-watt AM VHF Transmitter

The 1500 Transmitter is amplitude-modulated equipment intended for en route or airport service in the VHF airmobile band 118–136 mHz.

The equipment is fully solid state and meets ICAO recommendations for operation in 25kHz channelling environments.

SPECIFICATIONS

Frequency coverage:	118-136 mHz.
Number of channels:	One (External drive input available on rear panel)
Transmission mode:	A3 (DSB) up to 95% depth
Output power:	50 watts carrier (200W p.e.p.) into 50 ohm load impedance

Channel Scanning Monitor Receiver, Type 1700

The 1700 Channel Scanning Monitor is a multipurpose crystal-controlled 8-channel receiver operating in the VHF air mobile band 118-136 mHz AM.

The signal searching technique of the monitor is accomplished by the use of low current drain logic circuits which permit operation from internal rechargeable batteries (optional). An integral telescopic antenna is included which allows operation in both desk top or hand-carried modes.

Provision is made for the deselection of unwanted channels and the automatic mode may be inhibited to permit manual operation of any one channel.

SPECIFICATIONS

Frequency coverage:	118-136 mHz.
Number of channels:	Up to eight—no bandwidth restriction

1000 series UHF AM Receiver

The receiver is intended for rack mounting in airport or en route stations and may be operated from standard AC or 240V DC supplies. An auto no-break changeover facility is incorporated which will cause the receiver to switch to DC supplies if the AC mains fail.

SPECIFICATIONS

Frequency coverage:	225 - 400 mHz A3 amplitude modulation
Number of channels:	One
RF input impedance:	50 ohm unbalanced (N Type connector)

25W AM VHF Transmitter, 1250 Series

The 1250 Transmitter is amplitude modulated equipment intended for en route or airport service in the VHF airmobile band 118-136 mHz.

SPECIFICATIONS

Frequency coverage:	118-136 mHz.
Number of channels:	One (external drive input available on rear panel)
Transmission mode:	A3 (DSB) up to 95% depth
Output power:	25 watts carrier (100W p.e.p.) into 50 ohm load impedance

1400 Series 2-10 watt AM VHF Transmitter

The 1400 Transmitter is amplitude-modulated equipment intended for airport service in the VHF airmobile band 118-136 mHz.

The equipment is fully solid state and meets ICAO recommendations for operation in 25kHz channelling environments.

SPECIFICATIONS

Frequency coverage:	118-136 mHz.
Number of channels:	One (External drive input available on rear panel)
Transmission mode:	A3 (DSB) up to 95% depth
Output power:	10 watts carrier (40W p.e.p.) into 50 ohm load impedance. Simple internal adjustment gives any power output between 2-10 watts.

Line-Terminating Unit, Type 1350

The Type 1350 line terminating unit is designed for use with Park Air transmitter receiver type 1300. The unit is designed to provide remote control facilities for the 1300 equipment operating over two wire telephone lines.

Multichannel Receiver, Type 1900

The Park Air 1900 Receiver has been specially developed for use in conjunction with the Park Air 1230 Drive Units.

SPECIFICATIONS

Service:	A3 amplitude modulation
Frequency range:	118–136 mHz.
Channels:	720
Channel spacing	25 kHz

Receiver, Type 1800

The Park Air 1800 Receiver has been specially developed for use in conjunction with the Park Air 1210 Drive Unit and an associated transmitter type 1400/1250 or 1500. The receiver is crystal controlled with up to eight channels which are selected via the 1210 switching circuits. DC power for the receiver is obtained from the associated transmitter.

SPECIFICATIONS

Service:	A3 amplitude modulation
Frequency range:	118–136 mHz.
Channels:	Up to eight—no bandwidth restrictions
Channel spacing:	25 kHz to 50 kHz

Communications Company, COMCO, has a new series of VHF radios for vehicles and fixed stations. The model 727 is a two-way radio for airport vehicles, available in a desk console for ground station applications at airports, airlines, oil rigs. The station can be operated both locally and by remote control over a two-wire telephone line.

The COMCO 733 is designed for individuals who need to communicate directly with aircraft or control towers for safe, expeditious, and economical operation of aircraft. The model features hand-held design, four channels, unlimited spacing between channels, long battery life.

Communications Company (COMCO)
7811 Coral Way Suite 106
Miami, Florida 33155

DORNE AND MARGOLIN, INC.

An aircraft's radio is designed to deliver maximum performance by using an external antenna that is mounted on the aircraft so it delivers the RF signal to the radio. In addition, aircraft radios, tower transmitters and VOR stations use less than 100 watts of transmitted power. As a result, the antenna used on an aircraft is extremely important. Dorne and Margolin manufactures antennas that are designed to give a pilot maximum receiving and transmitting range for the aircraft's transceiver; obtain navigation accuracy, stability and range with the VOR; reduce noise caused by static; protect avionics from lightning strikes; and be weather and corrosion proof.

Dorne and Margolin, Inc.
2950 Veterans Memorial Highway
Bohemia, New York 11716

FOSTER AIR DATA SYSTEMS

Foster Air data produces several systems for area navigation:

The AD611 RNAV can be specified to include up to eleven waypoints or as few as one waypoint. The eleven-way system allows pilots to store an entire flight in the memory components and use the spare manual waypoint for impromptu waypoints during the flight. For single-pilot flight operations, manual waypoints can be added to the basic RNAV computer system to provide a one, two, or three waypoint system that is straightforward and easy to use.

Foster Air Data AD611 RNAV

SPECIFICATIONS

COMPONENT	SIZE	WEIGHT	POWER
61 RNC Remote Navigation Computer	12½" × 6-3.4" × 3¾"	4.5 lbs. 2.0 kg.	1.8 Amp @ 14VDC .9 Amp @ 29VDC
61 SIU System Interface Unit	3½" × 5½" × 6¼"	2.2 lbs. 1.0 kg.	from 61 RNC
61 DRM-1 Digital Range/Mode Selector—with groundspeed	*Q-Pack	0.8 lbs. 0.4 kg.	from 61 ARNC
61 DRM Digital Range/Mode Selector—less groundspeed	*Q-Pack	0.8 lbs. 0.4 kg.	from 6 RNC
61 WPS Waypoint Setter	*Q-Pack	0.8 lbs. 0.4 kg.	from 61 RNC
Memory Waypoint System (10 MWPS)			
61 HDU Horizontal Display Unit		0.8 lbs. 0.4 kg.	from 61 RNC
61 DEU Data Entry Unit		0.8 lbs. 0.4 kg.	from 61 RNC

*Standard Q-Pack is 3" × 1⅝" × 6½". Mounts separately. Regulatory Compliance: FAA AC90-45A

The RNAV 511 is a low-cost, multiwaypoint navigation aid for light aircraft. It will interface to most VOR/DME/CDI/HSI and autopilot systems in light aircraft. Equipped with an easy to use fault analysis diagnostic system, RNAV 511 computes to airports 199.9NM distant.

SPECIFICATIONS

	SIZE	WEIGHT	POWER
RNAV 511 PANEL UNIT	6¼" × 2⅜" × 8¾"	2.5 lbs.	1.2a @ 14dvc, 0.7a @ 28vdc.
RNAV 511-C PANEL UNIT	6¹¹⁄₁₆" × 2⅜" × 8¾"	2.5 lbs.	1.2a @ 14vdc 0.7a @ 28vdc
51DSA DELTA STEERING ADAPTER	6" × 4" × 1⅜"	1 lb.	supplied by panel units
51DSA/c328 DELTA STEERING ADAPTER	6" × 4" × 1⅜"	1 lb.	
51ASA ALPHA STEERING ADAPTER	6" × 4" × 1⅜"		

The VNAV 541 vertical navigation system provides altitude guidance advisories for cruise descents and nonprecision instrument approaches. During VFR cruise descent, it provides altitude guidance to the airport traffic pattern altitude, and plans a six-mile slowdown zone as preparation for entering the landing pattern.

VNAV 541 (TOP) and RNAV 511

SPECIFICATIONS

Weight:	0.7 lb.
Power Requirements:	.4 amps @ 4v,.2 amps at 28v
Altitude Accuracy:	±100′ (Dependent on RNAV DTW Accuracy)
Resolution:	100′
VNAV altitude display:	100′ increments to 65,000′
Altitude set:	100′ increments 0-5000′ 500′ increments 5000-18,000′ 1000′ increments 18,000-24,000′
MDA set:	100′ increments
Maximum computed altitude:	65,000′
Flight path angles:	1.5°, 3.0°
Signal Inputs:	1) Serial BCD Range to Waypoint data 2) Distance to Waypoint Range Block 3) 40mv/NM (requires external adapter)
Outlets:	Serial BCD VNAV altitude, selected altitude and mda
Environmental:	Temperature range 1.5°C to + 70°C altitude 50,000′
Display dimming:	Automatic via photocell
Mounting:	Front (surface) or Rear (flush)
Displays:	7 segment incandescent display, digits in 100's of feet 1.5° and 3.0° lighted pushbuttons

The RNAV 612 was designed to be the center of navigation information in cockpits of cabin class twins, turboprops, and jet aircraft. The four way-point design in combination with a new and exclusive automatic waypoint feature fills all enroute and terminal RNAV applications without need for mass memory waypoint storage. Equipped with its own integral VOR/localizer converter, RNAV 612 can be used in a new or existing aircraft.

Foster Air Data Systems
7020 Huntley Road
Columbus, Ohio 43229

GENERAL AVIATION ELECTRONICS

Genave produces radios for use by airports for primary as well as emergency communications, by airplane pilots and ground crews, and as portable and vehicle communications by federal and state governments. Following are several examples:

The 3W p.e.p. 118-135.975 mHz Air Com offers four channels and three watts p.e.p. output power. Batteries, antenna, and charger are included with the unit.

SPECIFICATIONS

Frequency range:	118.00 to 135.975 mHz 118-128 mHz standard
Number of channels:	4
Frequency spread:	10mHz
Channel separation:	25 kHz

The Alpha 12 VHF-AM transceiver 118.0-135.975 mHz has 12 channels, four watts RF output power, 25 kHz channel spacing, front panel headphone and microphone jacks, backlighted channel dial.

SPECIFICATIONS

Frequency range:	118.00-135.975 mHz.
Number of channels:	12
Channel separation:	25 kHz

The Alpha/720 transceiver allows for the transmission and reception of all 720 channels in the communication network (118.00-135.975 mHz). The Alpha/720 features four watts transmitter output power nominally and 16 watts peak.

SPECIFICATIONS

Frequency range:	118.00-135.975 mHz.
Number of channels:	720
Channel Separation:	25 kHz

The GA/1000 NAV/COM unit allows the pilot to make his own installation decision: a one-unit panel position or two units with a remote CDI location. It

Genave AirCom

Alpha/720 Transceiver

provides all 720 channels of VHF communications and 200 channels of VOR/LOC. Receivers are independent.

SPECIFICATIONS

Frequency range:	108.00-117.95 (Navigation) 118.000-135.975 (Communication)
Number of channels:	Navigation—200 Communication—720
Channel separation:	Navigation—50 kHz Communication—25 kHz

RECEIVE COMMUNICATIONS

Sensitivity:	2 µV max at 1 kHz, 30% AM, 6 db $\frac{S+N}{N}$
Selectivity:	−6 db ±8 kHz, −60 db ±25 kHz } 6 pole, 1db Cheb. filter, shape factor 2.5

VOR

Indicator needle stability:	Comm. transmitter shall cause no visible deflection of course deviation indicator needle.
Accuracy (normal & Reciprocal)	±2° at 90° points and ±3° at 45° from 10 µV to 20,000 µV signal.
Meter deflection	Full scale at 15°
Deflection sensitivity	At least ½ inch for 10° at µV to 20,000 µV.
Deflection linearity	10% proportional to phase.
Deflection response	Abrupt change shall reach 70% of ultimate position in 3 seconds with less than 20% overshoot.
Ambiguity flag	5 µV at 45° points, 3 µV at 90° points.
Sensitivity	From 10-20,000 µV input signal bearings up to ±60° from selected radial shall be clearly "TO" or "FROM". "OFF" if either 9960 or 30 Hz is missing. "OFF" if RF signal is missing.
Auto pilot output	150 mv for a full scale deflection (ARINC Standard).

LOCALIZER

Centering accuracy	±½ db 10 µV to 10,000 µV input signal.
Flag sensitivity	5 µV
Deflection sensitivity	20 µV should produce 60% of standard deflection at ±4 db.
Auto Pilot output:	150 mv for a full scale deflection (ARINC Standard).

The Alpha-Six is a six-channel VHF/AM aviation-band portable transceiver.

SPECIFICATIONS

Frequency range	118-136 mHz.
Number of channels	Six
Frequency spread	18 mHz

GA/1000 NAV/COM

Genave power supplies convert electrical voltage from 117 AC to 13.7 DC for more convenient operation of a Genave radio.

General Aviation Electronics, Inc.
4141 Kingman Drive
Indianapolis, Indiana 46226

GLOBAL NAVIGATION

Global Navigation pioneered in the late 1960s the practical use of Very Low Frequency (VLF) Naval Communications stations for aircraft navigation. The GNS-500A system is fully TSO'd and approved for IFR enroute RNAV operation in the continental United States, the District of Columbia and Alaska. It is also approved for navigation in the North Atlantic MNPS airspace. The system receives up to 16 stations of both the U.S. Naval Communications and Omega Navigation networks, providing worldwide navigation capability at all altitudes.

To operate, the pilot programs the latitude and longitude of the starting point and up to nine waypoints. The system then displays enroute information such as present position, track, ground speed, miles-to-go, and cross-track data. Standard ARINC outputs are provided to interface with flight director and autopilot systems.

Options are available to display desired track, bearing-to-waypoint and distance-to-go on the horizontal situation indicator.

Global Navigation
2144 Michelson
Irvine, California 92715

KING RADIO CORPORATION

King Radio produces aircraft communication, navigation, and light control equipment.

The King Silver Crown line of avionics is for single-engine and light-twin aircraft. Items include radar altimeters, audio control systems with marker beacon receivers, digital weather radars, navigation receivers, compass systems, digital DMEs, transponders, integrated NAV systems, integrated NAV systems with nine-waypoint capability, digital ADF, VHF COM transceivers, airborne radio telephones, NAV/COM systems, encoding altimeters, and blind encoders.

The King Gold Crown line of avionics is for business aircraft. Avionics include a choice of digital keyboard control or conventional mechanical frequency selectors, 10-waypoint storage area navigation system, integrated automatic flight control systems with vertical navigation, and radar altimeter.

The King Airline (ARINC) equipment features VOR/ILS receivers, digital DMEs, ATC transponders, VHF COM transceivers.

King Radio Corporation
400 N. Rogers Road
Olathe, Kansas 66061

MENTOR RADIO COMPANY

Mentor produces transceivers and audio panels for aircraft. Model TR-12 has the option of one to ten channels, between 118.000-135.975 mHz. To achieve low noise, sharp selectivity and low spurious responses, the design incorporates a four-pole monolithic crystal filter, a dual-gate field effect mixer and a triple-tuned r.f. amplifier. Squelch is controlled from the front panel.

Model TR-12 Portable COM transceiver has a rechargeable battery system and 115 VAC operation.

Model TR-12F Communications transceiver has all the same features and performance of the basic TR-12 plus ± .002 percent frequency tolerance to meet FCC requirements for fixed base transmitters.

The Mentor Audio Panel simplifies radio management by centralizing controls of multiple unit avionics systems. Receivers can be selected for speaker or phone output, either individually or in combination without audio loss due to mutual loading. The microphone can be switched to the desired transmitter or to a public address mode. An integral isolation network and solid-state amplifier assures full audio power for all radios. The control panel is

Model TR-12F Transceiver

APM-1 Audoi Panel

backlighted for night visibility. The model includes a complete three-light marker beacon receiver for both visual and audible indication of passage over these navigation aids. A Hi-Lo test switch provides two levels of marker receiver sensitivity and means to test the three-marker light circuits.

The Audio Panel comes in two models: AP-1 and APM-1.

Mentor Radio Company
1561 Lost Nation Road
Willoughby, Ohio 44094

DAYTON-GRANGER MARTECH

Dayton-Granger Martech manufactures VHF communication antennas, VOR V/Glide Slope antennas, navigation antennas, antenna couplers, VHF business communication antennas, UHF business communication antennas, cable kits, and adaptors.

The Omega Static Discharger has a lightning diverter element.

SPECIFICATIONS

P/N	MODEL NUMBER	MOUNTING RADIUS—"R"
16286	611R-4A	Flat
16293	611R-11A	5″ (12.7cm)
16290	611R-8A	3.5–4.81″(8.89–12.2cm)
16292	611R-10A	3.00–3.92″ (7.62–9.95cm)
16288	611R-6A	2.5″ (6.35cm)
16289	611R-7A	2–2.36″ (5.08–5.99cm)
16287	611R-5A	1.43–1.6″ (3.63–4.06cm)
16291	611R-9A	1.25–1.37″ (3.17–3.47cm)
16297	611R-15A	.689″ (1.75cm)
16296	611R-14A	.531″ (1.35cm)
16155	611R-16A	.5″ (1.27cm)
16348	180-12	.38″ (.96cm)

Dayton-Granger
P.O. Box 14070
812 N.W. First Street
Fort Lauderdale, Florida 33302

Martech
P.O. Box 1539
812 N.W. First Street
Fort Lauderdale, Florida 33302

MULTITECH

Sound equipment from Multitech includes the Boom Mike/Headset, model RS-300, which weighs less than one ounce, eliminates the need for headbands, and allows a pilot to use eyeglasses or an IFR hood without interference from auxiliary clip-on devices. The system also reduces cabin noises to a minimum. RS-300 unit includes microphone/earphone device, plug-in amplifier and retention clip for smaller ears.

An accessory item is the RN-10 Push-to-Talk Switch, which plugs into the amplifier or may be wired into the aircraft control panel.

For those pilots who only require a receiver, the TI-1 Earpiece fits over the ear, assuring clear sound. It can be used as a back-up receiver.

SPECIFICATIONS

RS300	
Microphone/Amplifier	
Connection	Microphone output: RCA phono plug Amplifier input: RCA phono jack Amplifier output: PJ-068 (.206″)
Cord length	5 feet
Supply current	15@ 13 volts
Impedance	Matches 100 to 600 ohms standard input
Type	Electret with acoustical tube
Output level	0.25 - 0.70 VRMS with normal voice and tube ¼ inch from lips. (Screwdriver adjust through hole in amplifier)
Receiver	
Frequency response	200 - 3000 Hz
Connection	PJ-055 (.250″) standard plug
Impedance	600 ohms
RN-10 SWITCH	
Cord	5 feet, coiled, shielded
Plug	Standard 3.5 mm
Mounting	Universal with adjustable nylon Fabcro-Lok. Fits any control wheel.
TI-1 HEADSET	
Frequency response	200–3000 Hz
Impedance	600 ohms
Net weight to wearer's ear	.7 ounce

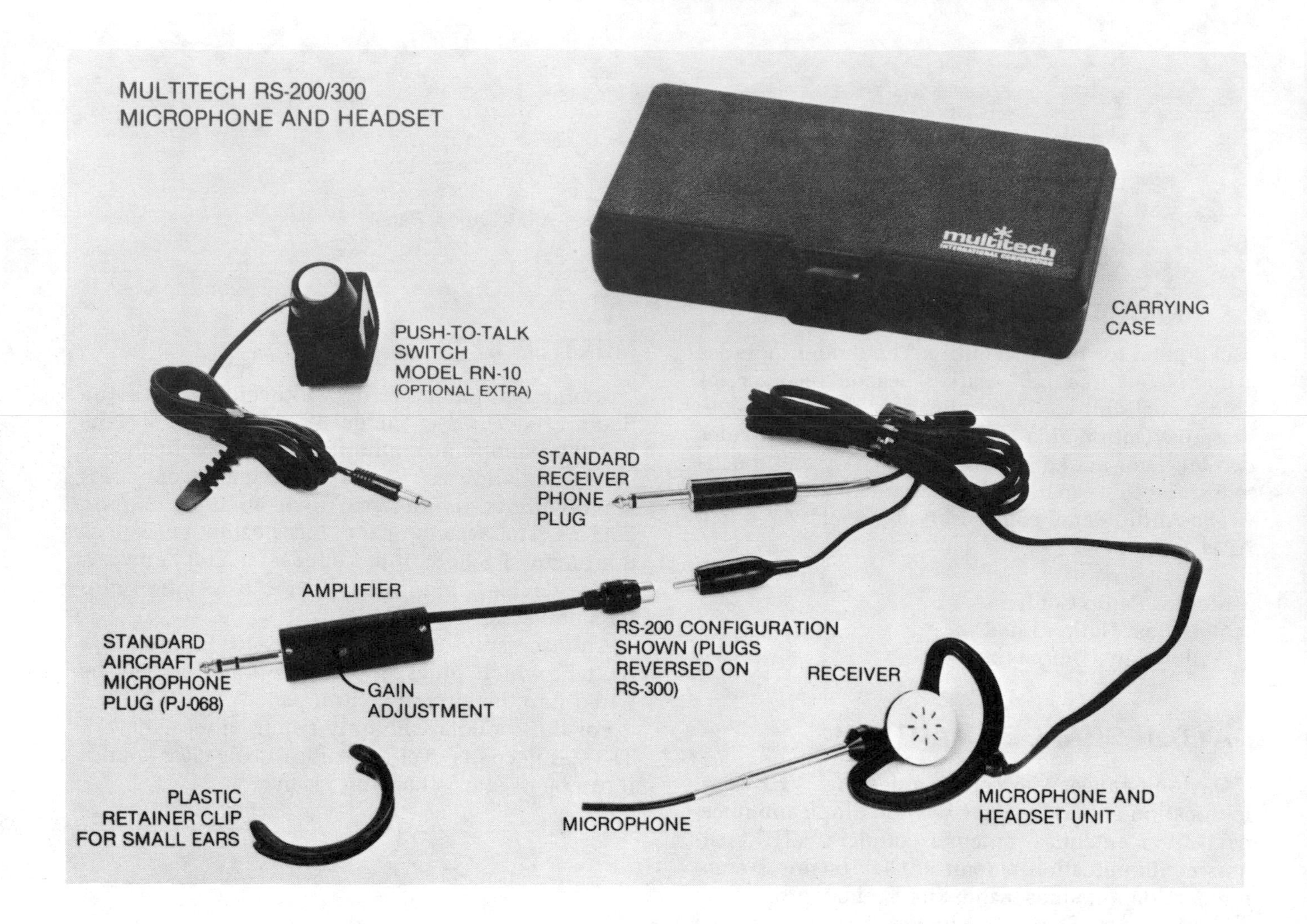

Multitech
8477 Enterprise Way
Suite 101
Oakland, California 94621

NORTON COMPANY

A properly designed radome allows the radar system to operate at maximum efficiency. Norton produces flush-mounted radome kits to accommodate the RCA Weather Scout 1 antenna on the Beech Bonanza aircraft. The kit is identified as P/N 4222X.

The radome kit for the Cessna 337 Skymaster, series 4231, is designed for flush mounting on the leading edge of the right or starboard wing.

Series 4229 for the Piper PA-32, 28 is also designed for flush mounting on the leading edge of the right or starboard wing.

Norton Company
P.O. Box 350
Akron, Ohio 44309

Cair Radome Kit 4222-X

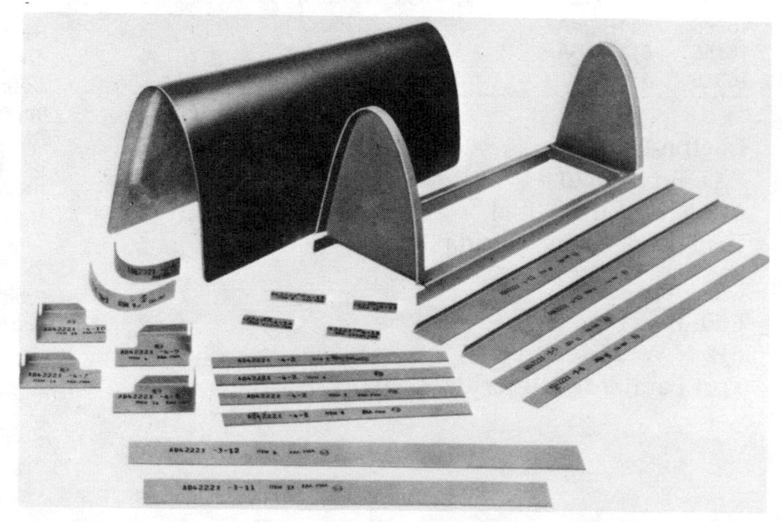

Kit# 4231X on Cessna 337

PLANTRONICS

The StarSet headset by Plantronics, for general aviation aircraft, eliminates the distractions of mikes and speakers in light aircraft communication systems. The microphone-receiver unit can be clipped to a lightweight headband or to a pilot's glasses. For aircraft without a built-in microphone switch, a push-to-talk switch is available. It mounts on the stick, permitting hands free operation. For an open cockpit or conditions where there is a lot of background noise, the noise suppressor, an optional feature, increases the signal to noise ratio. If two Plantronics push-to-talk switches are installed, the switch is designed so that the associated microphone is disconnected until that switch is activated.

Another general aviation headset from Plantronics is the MS series.

In addition, there are Plantronics headsets for commercial aircraft and also accessories, including

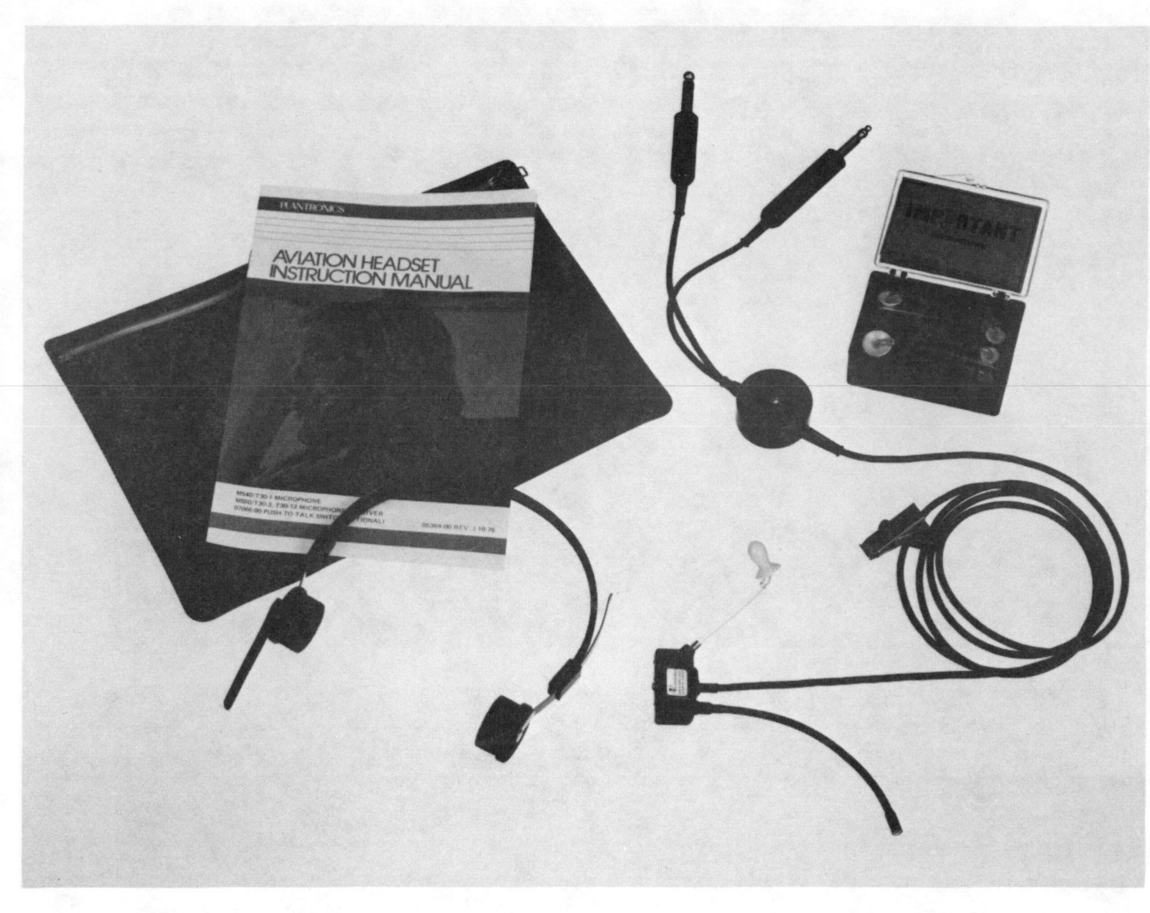

StarSet Headset by Plantronics

carrying case, push-to-talk switch, BNS to reduce cabin noise, background noise suppressor, eyeglass adapter, and headset pouch.

Plantronics
345 Encinal Street
Santa Cruz, California 95060

RADIO SYSTEMS TECHNOLOGY

Listed in the Radio Systems Technology catalog are audio panels, marker receivers, transceivers, unicom stations and battery packs, intercoms, lamp dimmers and antenna splitters, test equipment, headphones and microphones, antennas, sunscreens, and spare parts such as cable plugs and jacks. Some examples are:

RST-542 six-channel aircraft radio band transceiver with one channel's crystals.

RST-641 features a sturdy metal case, a six-amp-hour gelled electrolyte battery, jacks for microphone and headphone, voltage regulator, and wall charger. Additional options include internal speaker, PA capability, panel meter (battery voltage and signal strength), and up to three relays for remote activation of anything from runway lights to coffee pots.

RST-502 is an audio panel identical to the RST-503 except for the elimination of the marker beacon receiver circuitry. This allows the panel to be fabricated in half the chassis depth, permitting installation in aircraft with reduced panel room and shorter panel depth. Pushbutton switches selectively route audio signals to headphones and (if desired) the cabin speaker.

RST-443 intercom has four stations, includes background music input that is automatically overridden by any intercom or COM transceiver communication, toggle switch that allows the pilot in command to shut out passenger chatter if necessary during crucial transceiver communications and a relay circuit capable of providing voice-actuated operation of a small tape recorder. Options: radio interface and permanent mount.

RST-test equipment (models 601, 701, 711 and 721) is suited for servicing VOR/LOC, COM and MKR equipment. All RST test equipment can be powered from internal nine-volt batteries and all are provided with RF power output jacks on the rear panel. Units are portable.

Radio Systems Technology
10985 Grass Valley Avenue
Grass Valley, California 95945

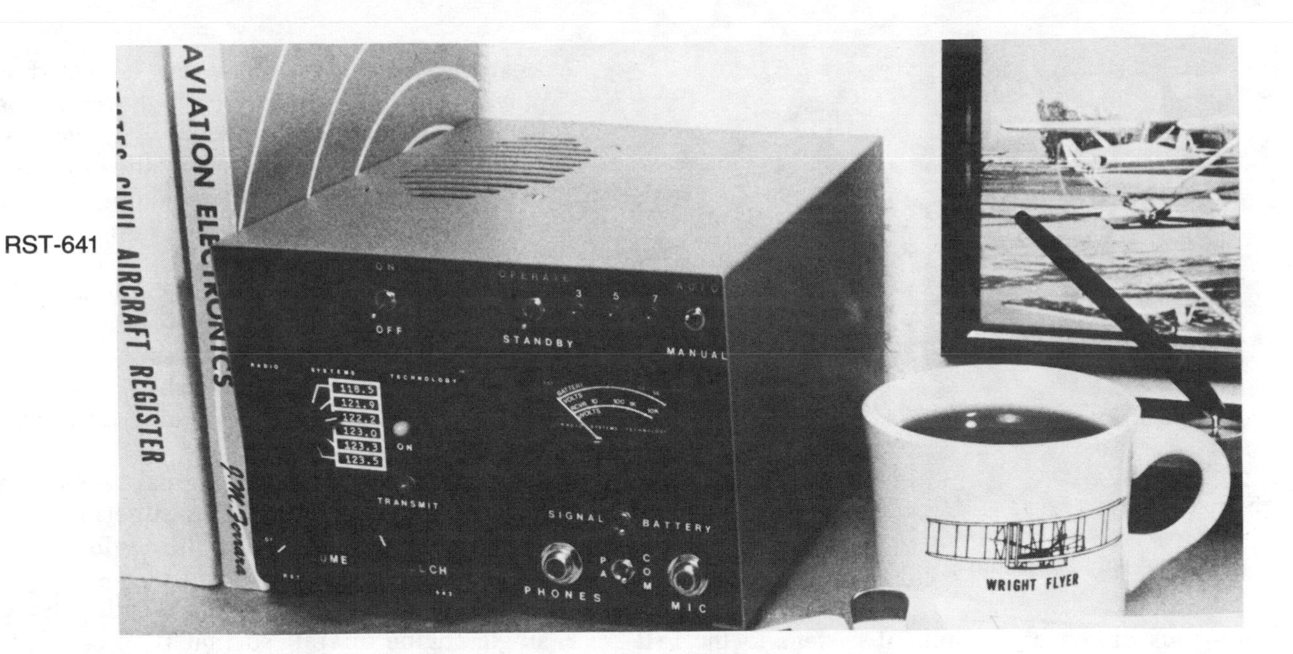

RST-641

SIGTRONICS

The Sigtronics' S-1M headset blocks out harmful cabin noise and provides clear, crisp communications. Volume control allows for adjustment for individual listening comfort and is helpful when mixing dissimilar headsets and hearing capabilities. High output 600 ohm dynamic receivers are efficient and provide good audio reception. A five-foot headset cord terminates in a "Y" with standard headphone and microphone plugs. Cushioned earcups accommodate individual head contours, and a padded headband adjusts for various head sizes. An amplified dynamic microphone was designed for use in high noise environments.

The transcom operates in any 12-volt fixed or rotary-wing aircraft. Twenty-four-volt units are available upon request. The unit is activated only when speaking, providing good noise reduction. The unit is voice activated: Pilot starts speaking and the transcom turns on to transmit the message to the other headset; pilot stops talking and it's off to reduce background noise. Transmitting can be accomplished from two positions. In addition, the hand-held microphone can be plugged into the transcom and used as an alternate mike. The intercom is automatically shut down during transmitting to allow only the voice of the person transmitting to go out over the air. The transmissions, however, are heard by both listening parties through the radio side-tone return. An audio cord provides radio monitoring capability, and radio output can be heard even with the transcom power switch in the "off" position. Power cord plugs into the aircraft cigarette lighter socket.

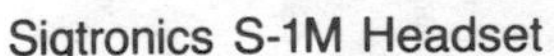
Sigtronics S-1M Headset

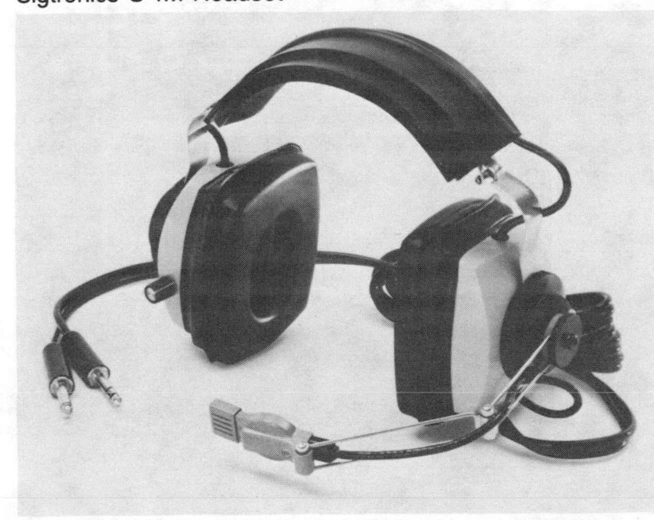

In addition to the portable unit, the transcom is available in a permanent mount and a remote mount unit. The permanent mount is identical in appearance and function to the portable unit with the exception of the power, headphone input, and mike output cords.

The four-way transcoms are functionally identical to the two-way units. Only the voice of the person transmitting goes out over the air because the other three microphones are disabled during transmit. The four-way units can be used with any number of headsets, because it is not necessary to have all four headsets plugged in for the unit to function. The portable model has a satellite unit with a four-foot cable that connects to the main unit as a two-way intercom.

The audio switcher enables the pilot and three passengers to listen to any programmed entertainment, but also be automatically switched from the

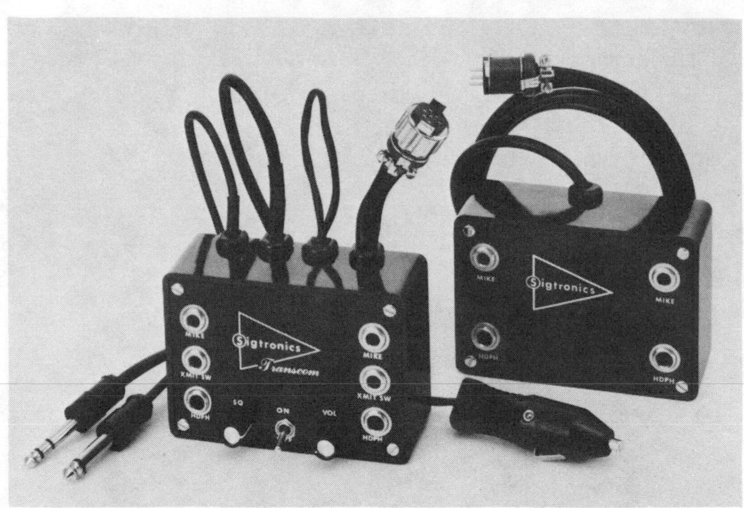

Sigtronics 4-way Transcom

audio bus of the entertainment system to the VHF radio when a radio message is received or transmitted. When the radio traffic is completed, the pilot and passengers are switched back to the entertainment.

The autocom operates in any fixed- or rotary-wing aircraft, helping to provide good, clear communications. Features include voice activation, self-contained battery enabling operation of the unit even in aircraft without electrical systems, operation with all standard aircraft headsets, radio monitoring, tape recorder output.

Sigtronics Corporation
824-A Dodsworth Avenue
Covina, California 91724

SIMULATOR FLYERS, INC.

The Hush-A-Com intercom for pilot and passengers eliminates the need for a hand-held microphone and is self-powered with an amplifier no larger than a pack of cigarettes. The unit includes two headsets with super noise-attenuating rings and noise-canceling microphones and a cable for direct monitoring of radio and power jack for 12–28 volt current.

For complete information, write to:

Simulator Flyers, Inc.
20 Church Street
Flemington, New Jersey 08822

SPERRY FLIGHT SYSTEMS

In 1976, RCA Avionics System, a part of Sperry Flight Systems, introduced ColoRadar to the aviation world, providing instant recognition of the weather situation with storms analyzed and displayed in three bright colors. Light rainfall is shown in bright green, medium rainfall in yellow, and a warning of heavy precipitation ahead in red.

The Primus-200 combines a color radar indicator with integrated receiver/transmitter/flat-plate antenna.

The Primus-300 Slim Line provides the advantages of a multicolor display with a 300-nautical-mile range. Among its features are positive-touch pushbutton controls, selectable azimuth marks, cyclic contour, target alert, and DataNav display capability.

The Primus-400, a 300 nautical mile digital radar, has an image rejection filter, superior antenna design, matched IF band and pulse width, AC cooling fan, and sector scan.

WeatherScout I scouts the weather for pilots of single-engine aircraft so that they can locate dangerous weather ahead. A radar system, WeatherScout I shows pilots storms and rain-related turbulence out to 90 miles ahead.

Primus-500 contains all function controls on the indicator panel. Other features include seven-color display, neutral density filter for clear and easy interpretation, mode annunciator with range identification, azimuth lines, and target alert.

Primus-100 has the receiver-transmitter-antenna in one unit. Pushbutton controls simplify changing ranges, modes, or functions. All function controls are located on the indicator panel.

WeatherScout I and II also scout the weather for

Primus 400 Indicator with Weather Display

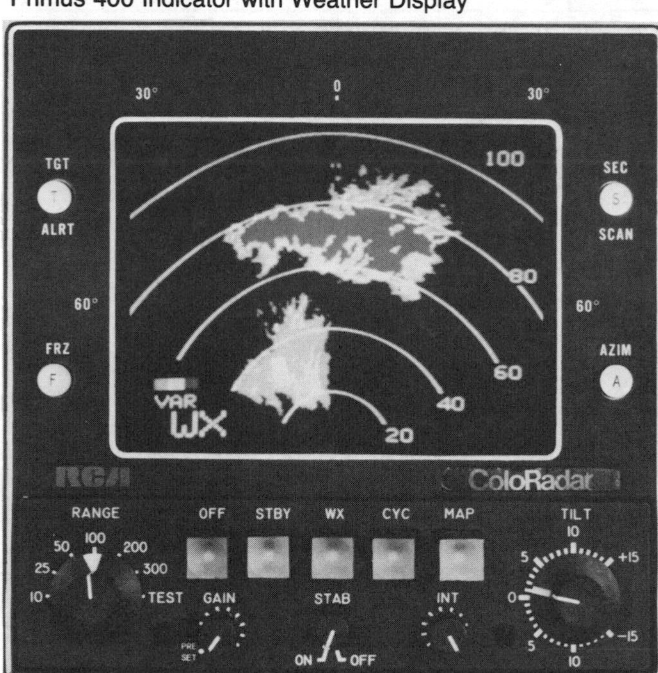

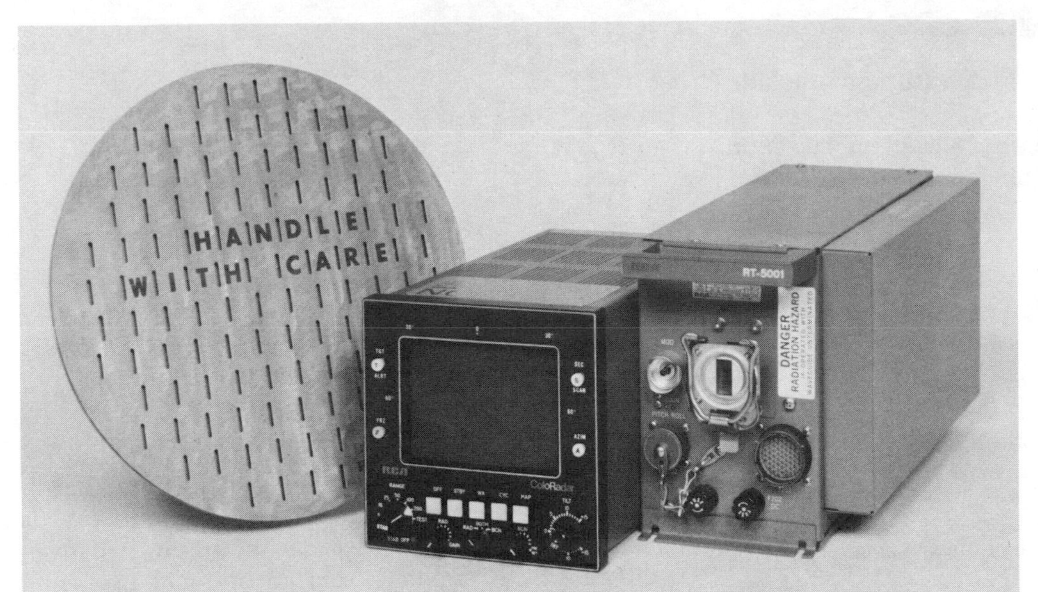

Primus 500 System

Primus 100 System

pilots, showing storms and rain-related turbulence within its 120 nautical mile range.

The 400 system allows a pilot flexibility in selecting the size of the indicator. All function controls are located on the indicator panel. Other features include easily accessible controls, bright six-color display, neutral density filter for bright, clear interpretation, mode annunciator with range identification, azimuth lines, and target alert.

Sperry Flight Systems
Avionics Division/Van Nuys
8500 Balboa Boulevard
P.O. Box 9028
Van Nuys, California 91409

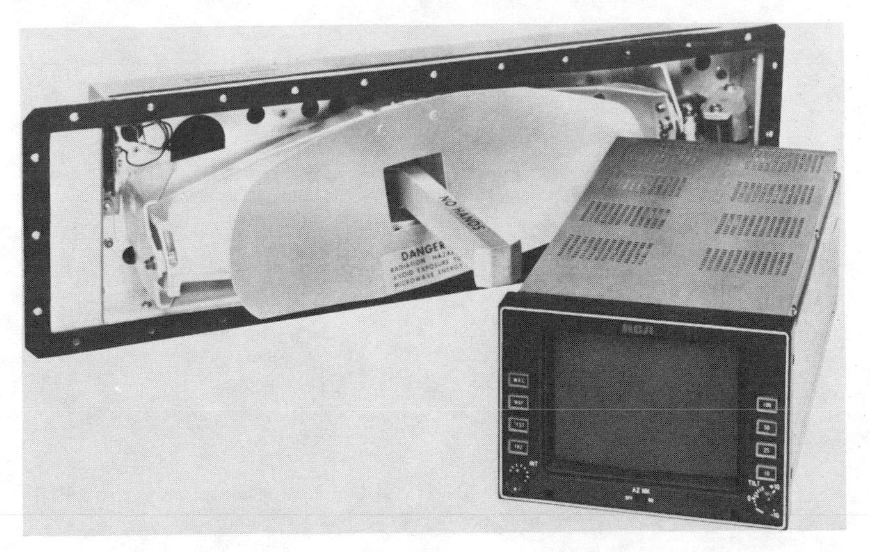

Color WeatherScout I System

SUPEREX ELECTRONICS CORP.

Communications products from Superex include the following:

APS II headphone is factory wired for 600 ohm impedance to match the phone output impedance of many communications receivers.

SPECIFICATIONS

Double headphone, monophonic wired	
Impedance:	600 ohms, or 8 ohm
Frequency Response:	100–7,000 Hz
Cushions:	Foam filled, vinyl jacket, replaceable, washable
Boom Microphone Mount:	Standard
Headband:	Vinyl covered spring steel with urethane foam cushion

APS-M3 headphone is factory wired for 600 ohm impedance, which can be changed to 4–16 ohm by the user, by moving a jumper wire located under the cushion.

SPECIFICATIONS

Double headphone, mono wired	
Frequency response:	100–9000 Hz
Impedance:	600 ohm; 4-16 ohm
Boom microphone mount:	Will accept Superex MFB and MRB boom microphones

APS-M3S: Single headphone version of APS-M3
Specifications: same as APS-M3

COM-2 is a lightweight headphone in which the frequency response is tailored for voice clarity.

SPECIFICATIONS

Double headphone, mono-wired	
Frequency response:	100–6,000 Hz
Impedance:	600 ohms, or 16 ohms (impedance must be specified at time of purchase)

COM2-S: Single version of COM-2
Specifications are similar and same options are available.

AVNS-7 is designed for use under very high ambient noise conditions, such as those found near jet aircraft or in high noise factory environments.

SPECIFICATIONS:

Double headphone, wired monophonic	
Dynamic driver:	600 ohms
Frequency response:	10–7000 Hz

Superex AVNS-7

M-710 base microphone features an omnidirectional condenser transducer with an FET pre-amplifier and a transistor amplifier. A unique feature is that the microphone stem can be unplugged from the base and operated from an optional extension cable.

Boom microphones are designed to be attached to Superex headphone models APSII, APS-M3, APS-M3S, COM-2S and AVNS-7. When purchased together with the headphone, the microphone and headphone cable can be supplied in one jacket. If not, microphones can be added to headsets later.

SPECIFICATIONS

MFB: Series of boom-mounted microphones feature a gooseneck style, flexible steel boom. A swivel base mounts this boom on the headphone.

MFB-2

Element:	600 ohms dynamic
Frequency response:	50-12,000 Hz
Mount:	Flexible gooseneck 6 inches long—fully adjustable. Attaches to Superex headphones listed above.

MFB-3

Element:	Ceramic element, 50K ohm Z
Frequency response:	200 HZ–8 KHZ

MFB-4

Element:	Carbon granule, 300 ohm z
Frequency response:	200–3500 Hz

MFB-5

Condensor/electret FET microphone	
Element:	Condensor electret with FET amplifier built in the microphone element.
Frequency response:	80 Hz–10,000 Hz
Require:	1½–9V power source for FET amplifier

Superex CEP-1 supply or LM5A amplifier will supply required voltage when used with MFB-5

The MRB series of boom-mounted microphones utilizes a rigid lightweight aluminum boom 11″ long. The position of the microphone is adjusted by sliding the aluminum boom in the mount and by rotating the mount.

MRB-2, MRB-3, MRB-4 and MRB-5, same as MFB-2, MFB-3, MFB-4 and MFB-5 respectively, except MRB series uses rigid aluminum boom.

All boom microphones except MFB-5 and MRB-5 are available in noise canceling design.

AM-3 microphone gets its power from an easily replaceable AA cell. The output impedance can be adjusted from 600 to 50,000 ohms.

LM-5 lapel microphone comes with a clip for attachment to the user's lapel.

ATM-5 is a condenser/FET microphone that is clipped to a light headband that holds the microphone element in place over the ear. A thin plastic sound tube located in front of the user's mouth conducts the voice up into the microphone chamber. The sound tube is acoustically tuned with the microphone front cavity to deliver a wide flat frequency response.

SPECIFICATIONS

Frequency response:	100 Hz–7,000 Hz
Weight:	2 oz. less cable
Cable:	6 ft. 3 conductor, no termination
Requires:	1½ to 6V power supply which can be provided by user. Can use Superex CEP-1 power supply, or use LM-5A amplifier, which will supply the required voltage, as well as 20 dB additional gain.

M-710 VOX, a voice-operated microphone, uses a condenser/FET microphone for input and uses integrated circuits. It has a .4 second time delay on switch-off to bridge syllable gaps. A push-to-talk paddle can override automatic operation. Microphone requires 12-volt DC external power source.

SPECIFICATIONS

Sensitivity:	−45 dB
Frequency response:	200–6000 HZ
Controls:	Trigger, power, PTT, Lock
Adjustment:	Volume

Lm-5A microphone amplifier can be used with low output microphones.

612 VX-VOX voice-operated electronic relay uses integrated circuits for sophistication and reliability. The microphone is amplified and frequency band limited, with a .4 second time delay on switch off to bridge syllable gaps. There is also a sensitivity control, and a provision for manual override.

- DPDT Relay output
- Requires −65 dB microphone input
- Requires 12 Volt 100 MA power supply

600 series stethoscope headphones feed sound to both ears through aluminum tubes.

SPECIFICATIONS (all models)

Frequency response	100–8000 Hz
Transducer	Dynamic

601 5 ft. cable, no termination

600-4 5 ft. cord—3.5mm plug termination

600-5 5 ft. cord ¼″ phone plug termination

600 1H-2000 ohm—5 ft. cable with ¼″ phone plug with matching transformer in plug barrel (barrel 2¾″ long)

Other items include VS-19 visor speaker; AV-720 headphone for general utility use with full-voice frequency response; S1-PTT push-to-talk switch for mobile applications with six-foot cable; F-S PTT push-to-talk switch in steel housing, designed for foot operation, and provided with eight feet of cable.

Superex Electronics Corp.
151 Ludlow Street
Yonkers, New York 10705

TELEX

Telex offers communication and navigations products, including headphones, microphones, full cushion headsets and lightweight headsets, slide computers, round computers, and plotters.

Telex 5x5 MARK IIA

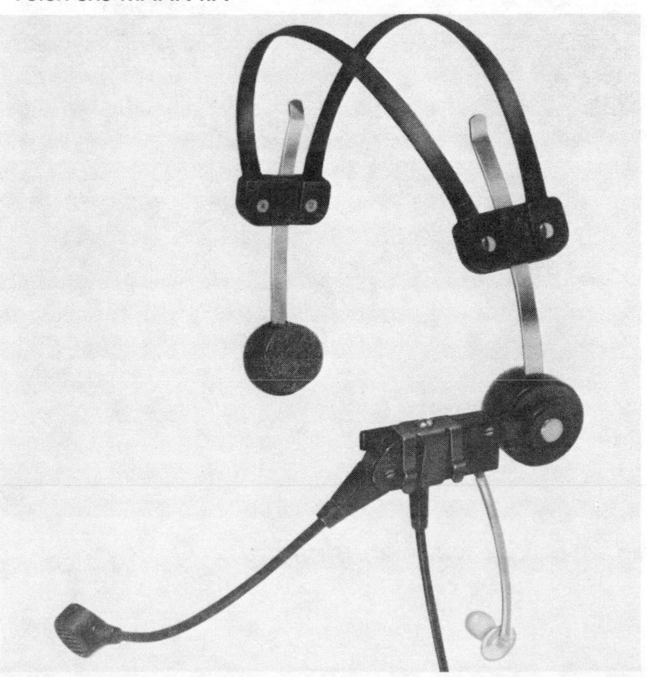

A few samples follow:

TELEX MODEL SELECTION GUIDE

F = FAIR G = GOOD E = EXCELLENT N.R. = NOT RECOMMENDED (Or characteristics not needed)

	OPEN COCK-PIT OR ROTARY WING	ENCLOSED SINGLE ENGINE COCKPIT-GENERAL AVIATION	LIGHT TWINS & BUSINESS AIRCRAFT	COMMERCIAL TURBO-PROP	COMMERCIAL JET
MICROPHONES					
66C	N.R.	F	F	F	F
66T	F	G	G	G	G
100 TRA	G	G	G	G	G
500T	E	E	E	E	E
HEADPHONES					
Twinset	N.R.	G	E	F	E
Pilot Earset	N.R.	F	G	F	E
A-610	F	F	F	F	F
A-1210	F	G	N.R.	G	N.R.
TH-900	E	E**	N.R.	E	N.R.
TAH-29	N.R.	G	G	G	G
HEADSETS*					
5×5 Pro II	N.R.	E	E	G	E
5×5 Pro I	N.R.	E	E	G	E
5×5 Mark II	N.R.	G	G	F	G
MRB 600	N.R.	F	G	F	F
MRB 2400	F	G	G	G	N.R.
DBM 1000	G	E	N.R.	E	N.R.
EBM 1400	E	E**	N.R.	E	N.R.

*Use with PT-200 push-to-talk switch on aircraft not so equipped.
**When ambient noise level exceeds 80 dB.

Telex
9600 Aldrich Avenue South
Minneapolis, Minnesota 55420

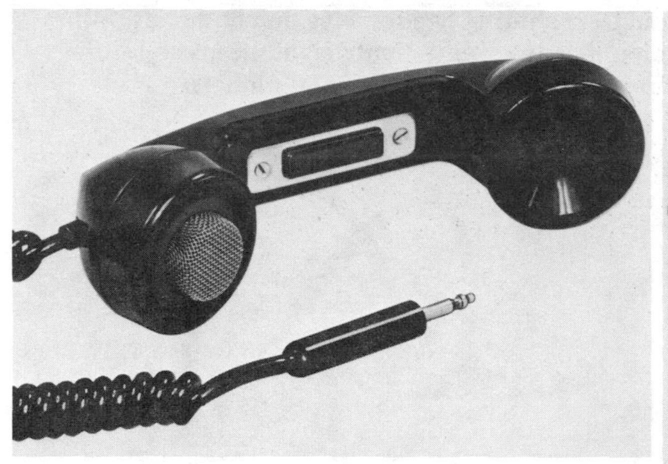

Telex HS-500 Handset

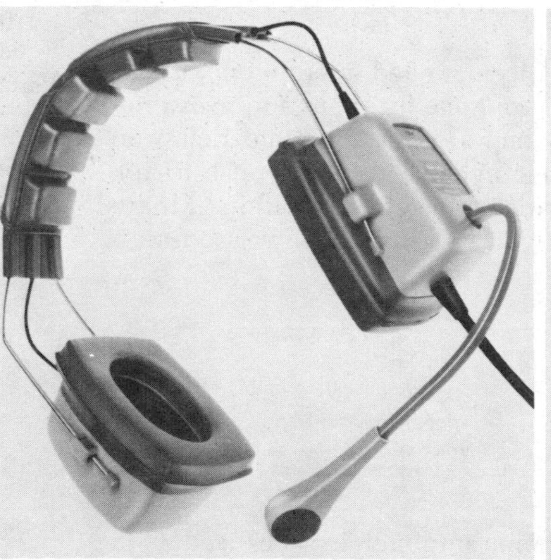

Telex DBM and EBM 1400 Hear-Defender

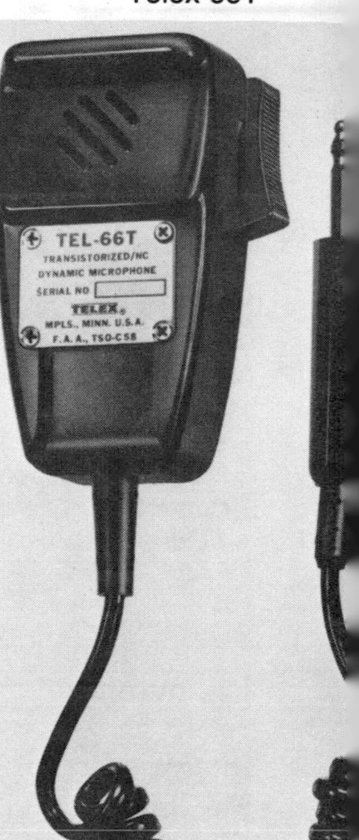

Telex 66T

7800 Omega/VLF Navigation System

TRACOR AEROSPACE GROUP

The 7800 Omega/VLF is a long-range navigation system. It is composed of three line-replaceable units (LRUs): the receiver/processor unit, the control display unit and the antenna coupler unit. The 7800 is certified for IFR enroute RNAV use in the U.S. and for sole means navigation in the Atlantic and Pacific.

Tracor Aerospace Group
6500 Tracor Lane
Austin, Texas 78721

REPAIRING YOUR AVIONICS EQUIPMENT

Even the best and newest avionics equipment may malfunction at times and require work. The following repair shops can help your new or old equipment. All are members of the Aircraft Electronics Association. Check with local aircraft manufacturers and distributors, with friends who are also interested in aviation, and scan your local telephone directory for other repair shops that may not be listed below.

ALABAMA

Birmingham: Hangar One Birmingham (205-591-6830).
Monroeville: Monroeville Aviation & Avionics (205-575-4235).
Montgomery: Montgomery Aviation, Dannelly Field (205-288-7334).

ALASKA

Anchorage: Aircraft Radio Service (907-272-7332); Alaskan Avionics Company (907-243-2370).

ARIZONA

Phoenix: Cutter Aviation, Inc. (602-273-1401); Sawyer Avionics (602-273-3770).
Scottsdale: Southwest Avionics (602-948-2400).
Tucson: Gates Learjet Corporation (602-746-5100).

ARKANSAS

Little Rock: Arkansas Modification Center, Inc. (501-372-1501); Central Flying Service, Inc., Adams Field (501-375-3245); Jet Corporation, Adams Field (501-372-5254); Hiegel Aviation, Inc. (501-375-9891).
Pine Bluff: Tomlinson Avionics, Inc. (501-534-0588).

CALIFORNIA

Burbank: Qualitron Aero, Inc. (213-843-8311).
Concord: Costal Avionics (415-687-3472).
Fresno: Frank X. Ruiz Electronics (209-233-0709).
LaVerne: Brackett Aircraft Radio (714-593-2596); American Pacific Avionics (714-593-3345).
Los Angeles: AiResearch Aviation Company, International Airport (213-646-5786).
Napa: Silverado Avionics, Inc. (707-255-5588).
Oakland: Tower Avionics Center (415-635-3500).
San Jose: L.A.C. Avionics, Reid-Hillview Airport (408-272-2458).
Santa Ana: Airadio Company (714-546-3325).
Santa Maria: Avionics West (805-925-6708).
Upland: California Air Radio, Cable Airport (714-985-0931).
Van Nuys: Beechcraft West (213-786-1410); IFR Avionics, Inc. (213-782-4810).

COLORADO

Broomfield: Avionics Associates (303-469-1707).
Denver: Aircraft Radio & Accessory, Stapleton Airfield (303-398-3717); Denver Avionics, Inc. (303-398-3883).
Englewood: Denver Jet Center (303-770-4321).

CONNECTICUT

Danbury: Danbury Aircraft Electronics, Inc., Danbury Municipal Airport (203-744-4408).
Stratford: Barbour-Daniel Electronics, Bridgeport Airport (203-375-5291).
Windsor Locks: Airkaman, Inc., Bradley International Airport (203-623-2671).

DELAWARE

Middletown: Summit Aviation, Inc., Summit Airport (302-834-5400).
Wilmington: Atlantic Aviation Corporation (302-322-7000).

FLORIDA

Clearwater: AeroComm Systems, St. Petersburg/Clearwater Airport (813-536-6565).
Ft. Lauderdale: Airborne Avionics Corporation (305-525-7403); Jet Executive International, Executive Airport (305-776-4781); Red Aircraft Sales, Inc. (305-523-9624).
Ft. Pierce: Ft. Pierce Flying Service (305-461-0600).
Jacksonville: Dumor Avionics, Jacksonville Inc., Jacksonville International Airport (904-757-2192).
Miami: Aero Systems Avionics Corporation (305-871-1300); Airport Radio Service, Inc. (305-526-6150); Peninsular Aircraft Radio (305-888-6713).
Opa Locka: Avionics Sales & Maintenance, Inc., Opa Locka Airport (305-592-8930); Dumor Avionics, Inc., Opa Locka Airport (305-685-3586).
Panama City: Sowell Aviation Company, Inc. (904-785-4325).
St. Augustine: Aerosport Avionics, Inc. (904-829-5707).
Sanford: C.E. Avionics, Sanford Airport (305-323-0200).
Tampa: Tampa Air Electronics Corporation (813-879-3713).
Vero Beach: Suntronics, Inc. (305-562-7900).
West Palm Beach: Butler Aviation, Palm Beach International Airport (305-683-5522).

GEORGIA

Atlanta: Avionics Sales Corporation (404-455-0348); Hangar One, Inc. (404-768-1000); Southern Aero Radio (404-691-5699).
Augusta: AiResearch Aviation (404-793-5600).
Macon: Electronic Craftsman Company, Lewis B. Wilson Airport (912-781-9940).
Peachtree City: Black's Custom Avionics (404-487-7903).
Savannah: Savannah Air Service, Inc. (912-964-1557).

ILLINOIS

Chicago: Butler Aviation International (312-686-7000).
Peoria: Byerly Avionics, Inc., Greater Peoria Airport (309-697-6305).
Polo: Radio Ranch, Inc., Radio Ranch Airport (815-946-2371).
Springfield: Capital Aviation, Capital Airport (217-544-3431).
Waukegan: Byrne Aviation Ltd. (312-336-8020).
West Chicago: JA Inc., DuPage County Airport (312-584-3200).
Wheeling: George J. Priester Aviation Service, Pal Waukee Airport (312-537-1200).
Wood River: Walston Aviation Sales, Inc. (618-259-3230).

INDIANA

Evansville: Ron Collins Aviation Electronics (812-425-2603).
Ft. Wayne: Ft. Wayne Avionics (219-747-1505).
Indianapolis: Combs Gates Indianapolis, Inc., Weir Cook Airport (317-243-3761); Sky Harbor, Inc. (317-293-4515).
Muncie: Muncie Aviation Corporation, Delaware County Airport (317-289-7141).
South Bend: Kerns Aircraft Electronics, Michiana Regional Airport (219-232-7965).

IOWA

Burlington: Electronic Applications Company, Inc., Municipal Airport (319-752-2435).
Cedar Rapids: Wathan Flying Service, Inc. (319-366-1891).
Des Moines: Des Moines Flying Shop (515-285-4221).
Spencer: Spencer Avionics, Municipal Airport (712-262-2364).

KANSAS

Olathe: Command Electronics Company, Johnson County Industrial Airport (913-782-7612); K.C. Piper Sales, Inc., Johnson County Executive Airport (913-782-0530).
Bevan-Rabell, Inc., Mid-Continent Airport (316-942-2461); Instruments & Flight Research, Inc. (316-684-5177).

KENTUCKY

Lexington: Vogt Avionics, Blue Grass Field (606-255-7004).

LOUISIANA

Baton Rouge: Louisiana Aircraft, Inc. (504-356-1401).

Lafayette: Air Logistics, Acadiana Regional Airport (318-233-1355); Paul Fouret Air Service, Inc. (318-237-0520).

New Orleans: Pan Air Corporation (504-245-1140).

MARYLAND

Glen Burnie: Baltimore Avionics (301-760-6669).

Hagerstown: Henson Aviation, Inc. (301-797-4100).

MASSACHUSETTS

Hyannis: Griffin Avionics, Inc., Barnstable Municipal Airport (617-771-2638).

North Adams: Sprague Aviation Company, Harriman Airport (413-663-5240).

Norwood: E.W. Wiggins Airways, Norwood Municipal Airport (617-762-5690).

Westfield: Northeastern Avionics, Inc., Barnes Municipal Airport (413-562-5124).

MICHIGAN

Adrian: Prentice Aircraft, Inc., Lenawee County Airport (517-265-8101).

Freeland: Air-Flite Serv-A-Plane, Inc., Tri-City Airport (517-695-2554).

Grand Rapids: Northern Air Service, Inc., Kent County Airport (616-949-5000).

Kalamazoo: Kal-Aero, Inc. (616-343-2548).

Lansing: General Aviation, Inc., Capital City Airport (517-321-7000).

Lapeer: G.B. DuPont Company, Inc. (313-664-6966).

Ypsilanti: Butler Aviation-Willow Run, Detroit-Willow Run Airport (313-482-2621); Quality Controlled Electronics, Inc., Willow Run Airport (313-485-4242).

MINNESOTA

Eden Prairie: American Aviation Company, Flying Cloud Field (612-941-4440).

Grand Rapids: Mesaba Aviation (218-326-6657).

Minneapolis: Northern Airmotive (612-726-5700).

MISSOURI

Chesterfield: Ellason Avionics, Inc. (314-532-3031).

St. Louis: Midcoast Aviation Service (314-426-7060); Rockwell International, Sabreliner Division (314-731-2260).

MONTANA

Belgrade: J&E Avionics (406-388-4161).

Billings: Aerotronics, Inc., Logan International Airport (406-259-5006).

NEBRASKA

Lincoln: Duncan Aviation, Inc. (402-475-2611); Lincoln Aviation, Inc., Municipal Airport (402-475-7621).

Omaha: Airkaman of Omaha, Inc. (402-422-6780).

NEVADA

Las Vegas: Hughes Aviation Services, Avionics Division (702-739-1100).

NEW HAMPSHIRE

Laconia: Furn Avionics, Inc., Laconia Airport (603-524-1515).

Manchester: Stead Avionics Company, Inc., Manchester Municipal Airport (603-669-4360).

NEW JERSEY

Newark: Butler Aviation-Newark, Inc., Newark Airport (201-961-2659).

Paramus: Butler Aviation International, Inc. (201-573-8000).

Teterboro: Teterboro Aircraft Service (201-288-1880).

NEW YORK

Buffalo: Metro Avionics, Inc., Greater Buffalo International Airport (716-632-1186).

Rochester: Page Airways, Inc. (716-328-2720).

Syracuse: SAIR Avionics (315-454-9221).

White Plains: Aircraft Electronics, Inc., Westchester County Airport (914-949-5375).

NORTH CAROLINA

Morrisville: Raleigh-Durham Aviation, Inc. (919-782-3232).

Rocky Mountain: Air-Care, Inc. (919-977-1717).

NORTH DAKOTA

Valley City: Flip's Avionics (701-845-2050).

OHIO

Akron: Laurence Aircraft Electronics, Inc. (216-733-7877).
Cincinnati: Avionics, Inc., Lunken Airport (513-871-6222).
Columbus: Capital Aircraft Electronics, Inc. (614-237-4271).
Elyria: Midwest Aviation, Inc. (216-323-7402).
Findlay: Marathon Oil Co., Aviation Division, Findlay Airport (419-422-2121).
Swanthon: National Flight Services, Inc. (419-865-2311).
Vandalia: Ohio Aviation (513-898-4646).

OKLAHOMA

Oklahoma City: AAR of Oklahoma, Inc. (405-681-2361); Air Center, Inc. (405-789-4020); Catlin Aviation Company (405-681-2331).
Tulsa: Autopilots Central, Inc. (918-836-6418).

OREGON

Troutdale: Western Skyways, Inc., Portland-Troutdale Airport (503-665-1181).

PENNSYLVANIA

Litiz: Lancaster Aviation, Inc., Lancaster Municipal Airport (717-569-5341).
Reading: Reading Aviation Service, Inc. (215-375-8551).

RHODE ISLAND

Westerly: North Eastern Electronics Company, State Airport (401-596-0735).

SOUTH CAROLINA

Aiken: Omnitronics, Inc., Aiken Municipal Airport (803-648-8895).
Charleston: Aero Aviation, Inc. (803-559-2401).
Spartanburg: Orr's Aero Mechnix, Inc., Downtown Airport (803-576-9442).

TENNESSEE

Chattanooga: Chattanooga Avionics, Lovell Field (615-894-1346).
Memphis: Aero Electronics, Inc. (901-345-2694); Avionics Specialist, Inc. (901-362-9700).

TEXAS

Addison: Executive Instruments, Inc. (214-239-0231).
Austin: Ragsdale Aviation, Inc. (512-926-7600).
Brownsville: Hunt Pan Am Distributors, International Airport (512-542-9111).
Dallas: Associated Radio Service Company (214-350-4111); Executive Aircraft Service, Inc. (214-357-1811); Foxtronics, Inc. (214-358-4425); K-C Aviation, Inc., Love Field (214-350-4177); Jet Fleet Corporation (214-350-4061); Red Bird Electronics, Inc., (214-337-8958); Rockwell International (214-996-5000).
El Paso: Avionics Associates (915-779-3481).
Fort Worth: Air Carrier Electronics, Inc., Meacham Field (817-626-2462); Metroplex Avionics, Inc., Meacham Field (817-625-4276).
Houston: Atlantic Aviation Corporation (713-644-6431); Jet Quarters (713-643-7059); Continental Radio, Inc. (713-644-1386); Jetronics, Inc. (713-641-0303); Woody Lesikar Aircraft Sales (713-492-2130); Temple Electronics Company (713-649-8175); Trunkline Gas Company, Houston Hobby Airport (713-644-1436).
Lubbock: Aero Communications, Inc. (806-765-6446); Texas Air Center (806-747-5101).
Midland: Aquila, Inc. (915-563-1300).
San Antonio: The Dee Howard Company (512-828-1341); Gen-Aero, Inc. (512-824-2313); Matthews Electronics, Inc., International Airport (512-826-2811).
Sequin: Horizon Communications & Electronics, Guadalupe County Airport (512-379-6705).

VERMONT

South Burlington: Northern Airways (802-658-2200).

WASHINGTON

Seattle: American Avionics & Instruments, Inc. (206-763-8530); Washington Avionics (206-762-0190).

WISCONSIN

Appleton: K-C Aviation, Inc., Outagamie Airport (414-735-2200).

Kenosha: Stateline Avionics (414-657-9331).
Milwaukee: Aviation Radio, Inc. (414-463-9180); Mitchell Aero, Inc. (414-747-5100).
Platteville: Wizard Radio (608-348-8514).

SAFETY AND EMERGENCY EQUIPMENT

AIRBORNE

The Ice/Breaker de-ice kits include de-ice boots plus dependable dry air pumps and de-ice controls. The kit operates by inflating tubes on the leading edge of the wings upon which the ice forms. As the tubes expand, the ice deforms and breaks and is then carried away in the airstream.

Airborne
1160 Center Road
Avon, Ohio 44011

ARP INDUSTRIES

Before the invention of equipment such as the Carburetor Ice Detector, a pilot might not have known that his carburetor was producing ice until the engine began showing problems. The detector detects the formation of frost and ice by using a system of blocked light rays. If a red light comes on during flight, the pilot merely applies carburetor heat until the red light signal stops. The light will come on up to five minutes before enough ice forms in the carburetor to cause loss of rpm and/or engine roughness. An optional audio warning is practical when a pilot's attention is distracted by cockpit chores.

The two models are a self-contained one, #107AP, which is panel mounted, and a remote one, #107AP-R, which is merely mounted.

A carburetor ice detector will solve other aircraft problems, including avoiding unnecessary use of

ARP Industries Model 107AP (TOP) and Model 107AP-R

carburetor heat, improving fuel efficiency, and extending engine life.

ARP Industries
36 Bay Drive
E. Huntington, New York 11743

AUTOTEC, INC.

Autotec manufactures an aircraft preheater and hydraulic ground support units.

The Mini-Pak I is a hand pump with integrated hydraulic manifold. It has a one-gallon capacity, a large filler cap with filtered vent ports and internal filler screen. It weighs only 15 pounds.

The Mini-Pak II, also a hand pump with integrated hydraulic manifold and one-gallon capacity, features not one but two independent hydraulic

Mini-Pak II

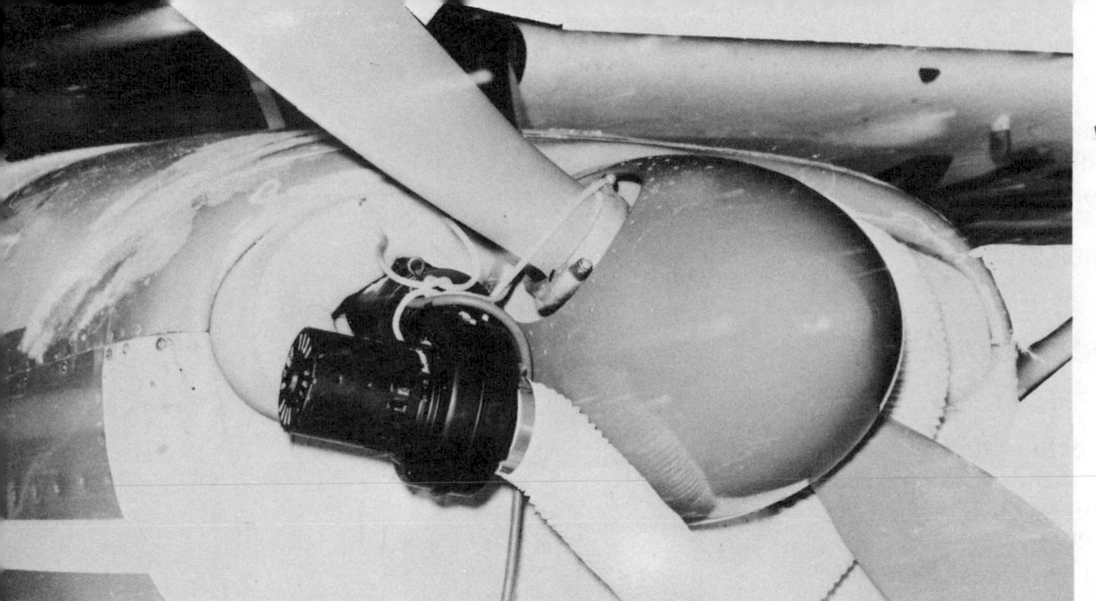

Warm-Start 150

systems in one compact unit. System 1 circulates hydraulic fluid from the plane's reservoir for accurate testing of all systems in aircraft. All hydraulic fluid passes through the cartridge filter. In system 2, hydraulic fluid is supplied from the Power-Pak reservoir. Fluid can be purged, filtered or added to the plane's system from the Power-Pak II reservoir.

The Warm-Start 150 preheater comes with built-in thermal overload, air sealing pads, hoses, clamps, lanyard, and instructions. The unit can heat engines or cabins and defrost windshields.

Autotec, Inc.
6555 Convent Blvd.
P.O. Box 391
Sylvania, Ohio 43560

AWK Universal Checklist II

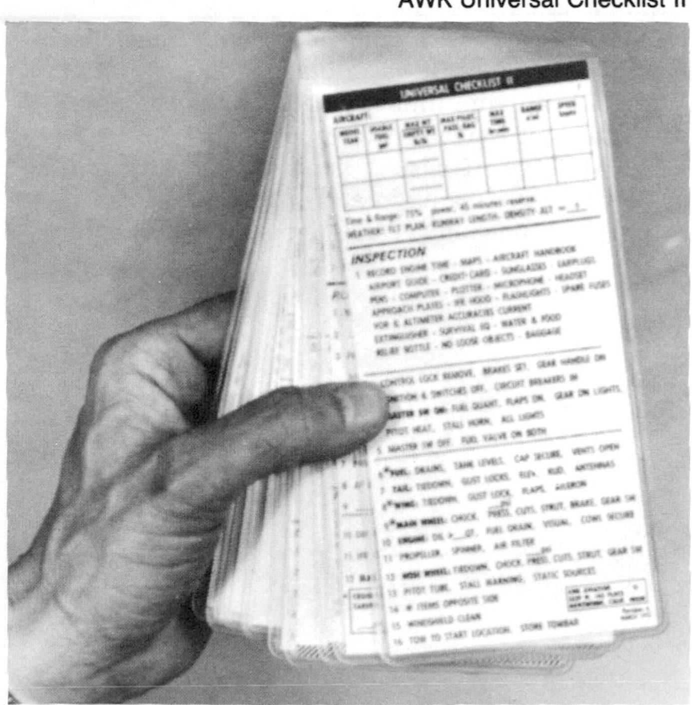

BEFORE START BONANZA V35B, F33A, A36 2

1 MAPS, IFR MATERIALS, FLASHLIGHTS ?

2 DOORS LOCK, SEAT ADJ, LOCK. SEATBELT, HARNESS.

3 FLIGHT CONTROLS FREE, CORRECT DIRECTIONS ?

4 ELECT EQ OFF. **BATT, ALT*SW ON.** GEAR LIGHTS GREEN ?
*IF EXT PWR: ALT SW OFF

5 **RADIOS:** ATIS COPY. GROUND, TOWER SET. VOR, ADF SET.

MARKER BEACON LIGHTS TEST. ELT ARM, INACTIVE ?

IFR -- CLEARANCE REQUEST

6 RADIOS, AUTOPILOT OFF. CLOCK SET, RUNS ?

7 FLAPS UP. ELEV, RUD, AIL TRIMS TAKEOFF. ELECT ELEV TRIM SW OFF

8 ALTIMETER = FLD ELEV ? AIRSPEED, CLIMB RATE = 0 ?

9 ALTERNATE STATIC SOURCE CLOSED

START ENGINE

1 BATT, ALT SW RE-CK ON. ROTATING BEACON ON

2 COWL FLAPS OPEN. PROP HI RPM

3 THROTTLE OPEN FULL

4 MIXTURE RICH

5 FUEL VALVE RIGHT. FUEL PUMP: ON - PEAK FLOW - OFF

6 THROTTLE OPEN ¼ IN.

7 **BRAKES ON, PROP "CLEAR ?"**

8 MAGS, STARTER ON

IF OVERPRIME:
a. MIXTURE OFF
b. THROTTLE OPEN FULL
c. MAGS, STARTER ON
d. THROTTLE IDLE, MIX RICH

9 IF HOT START: FUEL PUMP ON ---- OFF

10 1000 RPM WARMUP. **OIL PRESS ?**

11 IF EXT PWR: DISCONNECT, ALT SW ON ©

Universal Checklist II
SAMPLE PAGE 2 · MARCH 77
Filled in for Bonanza

AWK AVIATION
5539 WEST 142 PLACE
HAWTHORNE, CALIF. 90250

AWK AVIATION

The purpose of Awk's Universal Checklist II is to increase flight safety and efficiency by providing all of the essential items for the various phases of flight in a well-organized and complete form. Many types of light single-engine aircraft can be flown using the same checklist. The pilot only has to print in the airspeeds, flap positions, rpm's manifold pressures, and fuel/weight limitations. All phases of flight are covered in the checklist.

Awk Aviation
Universal Checklist
5539 West 142 Place
Hawthorne, California 90250

BUTLER PARACHUTE SYSTEMS, INC.

The Butler Parachute Systems Model BETA is an emergency parachute system designed from the ground up to use military surplus canopies in a lightweight backpack. The BETA is FAA approved in the Standard Category (high speed) in each of its standard variations.

Every Model BETA Parachute System has these features:

- MIL-Spec hardware, all new.
- Double main lift web harness design.
- Leg straps that are a continuation of the front risers and run through the main lift web without interruption.
- Entire container is constructed of double layer nylon parapack.
- Internal flaps of container are arranged to provide a staged deployment sequence by holding the canopy in place until the pilot chute gets a solid bite of air and pulls the canopy free.
- BETA has fully enclosed, shaped corners on the container to keep the canopy securely tucked inside, protected from the elements.
- BETA is available with the MA-1 pilot chute or the optional Hot Dog pilot chute which has larger spring.
- The ripcord housing is protected with a parapack cover that keeps it securely tucked in against the pack to prevent snagging.

Continental Air Sports
113 S. Monroe Siding Rd.
Xenia, Ohio 45385

McElfish Parachute Service
2615 Love Field Drive
Dallas, Texas 75235

Midwest Parachute Service
22799 Heslip Drive
Novi, Michigan 48050

Para-Gear Equipment Co.
3839 Oakton
Skokie, Illinois 60076

The RW Shop
Route 13
Brookline, NH 03033

Joe Smith Parachutes
P.O. Box 39
Lewisberry, PA 17399

Skydivers of Texas, Inc.
2553 Valley View Lane
Dallas, Texas 75234

DALE & ASSOCIATES

The Emeco Division of Dale & Associates sells a preheater. The blower unit is set in the window of a car, the window is rolled up until the unit is held securely, the magnetic vinyl shroud is pressed to the cap, and the plugs are stuffed into the aircraft inlet eyes. Finally the cord is plugged into the cigarette lighter. This entire procedure takes about three minutes.

Although the primary use of the preheater is to preheat the plane's engine, it can also be used to preheat the cabin for windshield frost.

Dale & Associates
1401 Cranston Road
Beloit, Wisconsin 53511

DEVORE AVIATION CORPORATION

By floodlighting its broad expanse, vertical tail lighting makes the flight path and proximity of an aircraft readily apparent to pilots of other aircraft, whether used in conjunction with, or to the exclusion of, other lighting.

Write for a list.

DeVore Aviation Corporation
16160 Stagg Street
Van Nuys, California 91406

DME CORPORATION

DME produces safety and survival equipment for the transportation industry. Products include:

Door-mounted lighting system, EL-104. High intensity flood light, illuminates deployed evacuation

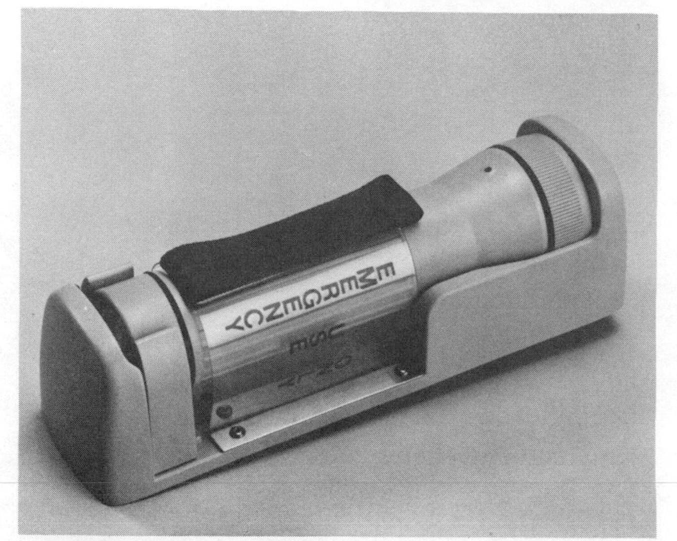

Portable Emergency Light EF-1

slide, automatically activated when door is opened, powered by own battery pack, weighs only 16 ounces.

Slide-mounted lighting system, ESL-1. Provides visibility during emergency evacuations, can increase evacuation rate, easily installed on existing slides or slide/rafts, automatically activated when slide is deployed, powered by long service life (four-year) battery pack, weighs 26 ounces.

Portable emergency light, EF-1 (photograph). High intensity hand-held light, immediately available during emergency, two to four times illumination of conventional flashlight, removal from bracket automatically activates unit, easily installed, no external electrical power supply needed, special battery pack provides 2.4 years service life.

Survival, utility, and first aid kit. All kits in waterproof packages, include treated drinking water, sea dye markers, knife, flares, raft repair kit.

Rescue locator radio beacons, RLB-9. Transmits simultaneously on 121.5 and 243.0 mHz, automatically activated when raft is inflated in water, manual activation on land, service life of 3.5 years provided by special battery pack, operational life of battery is 48 hours, weighs 31 ounces.

Plastic vacuum forming and formed metal parts. Product line also includes life raft canopy support rods.

Aircraft seat protective strips. Aluminum-mounted plastic strips protect them from damage.

DME Corporation
1631 S. Old Dixie Highway Building E
Pompano Beach, Florida 33060

EMERGENCY BEACON CORP.

Call for help at the flick of a switch using emergency devices from Emergency Beacon Corporation:

The Portable Rescue Beacon gives 72 hours continuous transmission on single replaceable battery, operates at −50° F to +160° F and is fully waterproof. Model No. EBC-102.

RATINGS

FAA TSO'd C61a
FCC Type Accepted Part 87 (25 kHz)
Meets all specifications International Civil Aviation Organization (ICAO) for personnel rescue beacons.
Meets all U.S. and Canadian specifications for personnel rescue beacons.

SPECIFICATIONS

Frequency:	121.5 mHz and 243.0 mHz simultaneously
Range:	To 150 miles
Operating life:	72 hours @ 50° F, 24 Hrs. @ 28° F, 15 Hrs @ −4° F
Operating temperature:	−50° F to 160°F
Modulation:	Swept tone 1000 Hz to 300 Hz, 2–3 times per second.
Power:	150 Mw
Submersion:	Withstands 20 hours min. complete submersion in salt water.

Portable Rescue Beacon

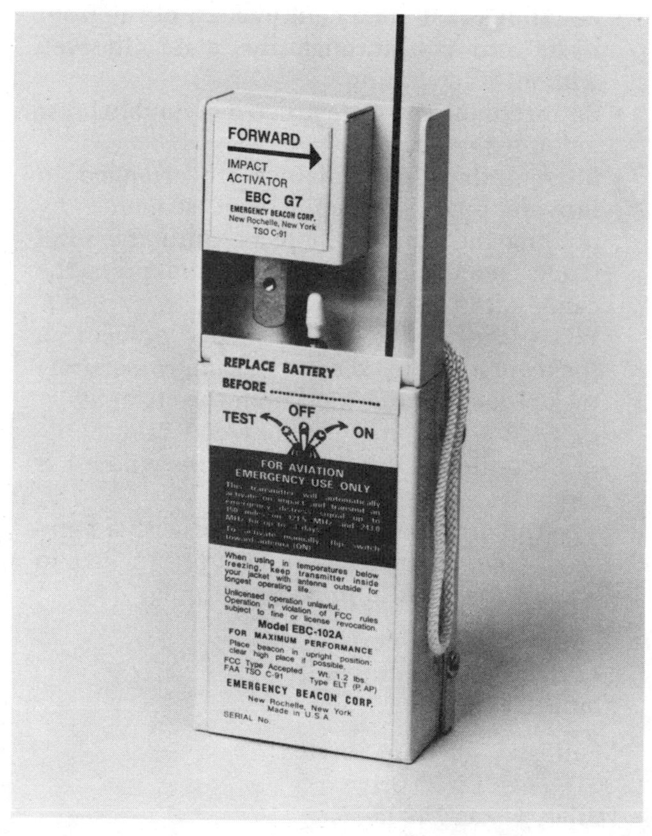

The Emergency Locator Transmitter, model EBC-102A, transmits for up to three days in all environments from arctic temperatures of −50° F to desert temperatures of +150° F without loss of range. It has solid state circuitry encapsuled in plastic for complete waterproofing and shock resistance, and comes with hand-hold lanyard and mounting kit.

SPECIFICATIONS

Range:	Up to 150 miles
Operating life:	72 hrs.
Battery:	Alkaline
Frequency:	121.5 mHz and 243.0 mHz simultaneously
Modulation:	Swept Tone 1000 Hz to 300 Hz 2–3 times per second
Shock:	Withstands 1000 G's
Waterproof:	20 hrs. complete submersion in salt water

PHYSICAL

Size:	2-½″ x 1¼″ x 6½″
Weight:	1.2 lbs.
Case:	Aluminum
Finish:	Yellow baked enamel Flexible non-telescoping antenna

The Emergency Locator Transmitter, model EBC 302, is both automatic and manual. In case of a crash, even if the pilot is injured, the built-in computer turns on the emergency beacon. It analyzes the output signal from the omnidirectional accelerometer and determines the nature of the shock. A built-in test light checks power output, modulation and battery.

The EBC 302V model has all the features of the EBC 302 and also allows a pilot to talk directly with the search aircraft and notify them of immediate needs, such as medical supplies, stretcher, or doctor.

SPECIFICATIONS

Range:	Up to 300 miles
Operating life:	200 hrs.
Battery:	Alkaline
Frequency:	121.5 mHz and 243.0 mHz simultaneously
Modulation:	Swept tone 1000 Hz to 300 Hz 2–3 times per second
Waterproof:	20 hrs. complete submersion in salt water

The Emergency Locator Transmitter, model EBC 302V, offers the following capabilities. It has solid-state circuitry encapsuled in plastic for complete waterproofing and shock resistance and comes with hand-hold lanyard and mounting kit.

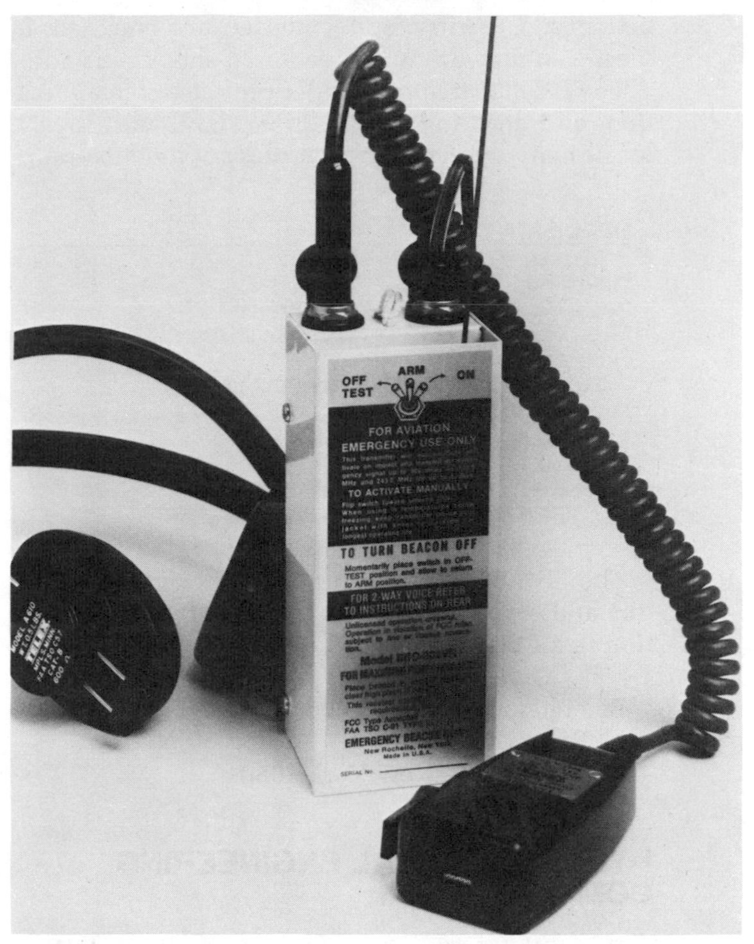

EBC 302V Beacon

SPECIFICATIONS

Range:	Up to 300 miles
Operating life:	200 hrs.
Battery:	Alkaline
Frequency:	121.5 mHz and 243.0 mHz simultaneously
Modulation:	Swept tone 1000 Hz to 300 Hz 2–3 times per second
Waterproof:	20 hrs. complete submersion in salt water

PHYSICAL

Size:	2½″ × 2″ × 6½″
Weight:	1.9 lbs.
Case:	Aluminum
Finish:	Yellow baked enamel Flexible nontelescoping antenna

The Automatic Emergency Locator Transmitter with voice transmitter-receiver feature, model 302 VR, is both automatic and manual and is equipped with a computer override fingertip switch to manually activate the emergency beacon. In addition, this transmitter draws power from two sets of batteries. Circuitry and batteries are enclosed in plastic so they are waterproof and shock resistant.

The DF-88 Direction Finder provides communication and locating capabilities simultaneously. It locates any transmitter, regardless of modulation.

SPECIFICATIONS

Frequency range:	118 mHz to 136 mHz or 238 mHz to 260 mHz
Supply voltage:	12 to 30 volts DC @ .045A
Antenna scanning rate:	75 times each second
Input impedance:	Greater than 1 megohm

The EBC-BR-15 Airborne Monitor Receiver monitors and responds with an aural and visual indication to an ELT signal TSO'd to C-91 on 121.5 mHz.

Emergency Beacon Corporation
13 River Street
New Rochelle, New York 10801

HYDRO-CHEMICAL ENGINEERING COMPANY

Granville's aircraft strut sealant stops leaking struts and also prevents corrosion. It coats the surfaces of the metal, forming an oil-retentive film that prevents oxygen transfer to the metal and thereby stops surface oxidation.

Strut seal leaks can be stopped by an owner or pilot in minutes, in just three steps:

1. Remove valve core, located at top of strut.
2. Insert plastic hose of empty container into valve core, three to five inches into strut. Depress plastic bottle and release to draw out enough fluid to raise level in bottle to line on container. This removes the proper amount of the fluid, which is then replaced with Granville's aircraft strut sealant.
3. Insert Granville's sealant tube into valve core approximately one-half inch. In normal upright position, press in sides of plastic bottle. When fluid is down to first line, you have the proper amount in strut. Top off with fluid removed prior to adding the sealant. Replace valve core. Air up. Expect to air a second time. Allow five to seven days to make complete seal.

Hydro-Chemical Engineering Company
1020 North Main
Morton, Texas 79346

INTERTECH AVIATION SERVICES

Intertech manufactures Res-Q-Pak Survival Systems for two or more persons, designed to help save and sustain lives in a downed aircraft situation. The systems are packaged in lightweight backpacks to further increase their usefulness and accessibility. Both are also equipped with heavy-duty signals and tools. The kits are available in three versions.

Intertech also manufactures a Halcon 1211 fire extinguisher, effective for small cabin fires and engine start-up fires and a downed pilot locator for pocket or flight bag.

RES-Q-PAC Type C

LOCATOR RATINGS

	FAA TSO'd C61a
	FCC Type Accepted Part 87 (25 kHz)
	Meets all specifications International Civil Aviation Organization (ICAO) for personnel rescue beacons. Meets all U.S. and Canadian specifications for personnel rescue beacons.

SPECIFICATIONS

Frequency:	121.5 mHz and 243.0 mHz simultaneously
Range:	To 150 miles
Operating life:	72 hours @ 50° F, 24 hrs @ 28° F, 15 hrs @ − 4° F
Modulation:	Swept tone 1000 Hz to 300 Hz, 2–3 times per second.
Power:	150 Mw
Submersion:	Withstands 20 hours min. complete submersion in salt water.

Intertech also manufactures supplementary and replacement items, including: strobelite, aerial flare, smoke bomb, sun signaling mirror, medical emergency kit, flashlite with filters, chemical light, rescue blanket.

Intertech Aviation Services
3 Sunset Lane
Littleton, Colorado 80121

MARTECH AND DAYTON-GRANGER

The Transceiver Series 150 from Martech provides emergency communications and can be stowed in general aviation cockpits until needed. Solid-state units, they provide two-way communications. The transceivers operate on VHF-AM emergency frequencies 121.5 or 123.1 MHz; the VHF emergency channel and emergency "scene of action" frequencies, respectively. The transceivers can be turned on by pilot or other crew member or passenger pulling the antenna out to its full length—approximately 24 inches. In rescue operations, these transceivers can be paired with emergency rescue beacons to provide voice communication between survivors and the rescue aircraft. There is both mercury battery model and water-activated battery model.

The PS-835 emergency power supply is a 25-VDC power system designed to drive self-contained gyro horizon indicators. It is also designed to accommodate a modular removable DC to AC inverter. If desired, the 25 VDC power output can be utilized for other equipment, such as an emergency transceiver.

J.E.T. also produces several inverters. The model S1-3003 is a three-phase inverter with fully regulated output power maintained with inputs from 24 to 32 VDC and at least 95 vac per phase under full load is maintained with inputs as low as 20 VDC.

Model S1-1500A is a TSO-approved, single-phase, solid-state inverter for new applications or to replace high maintenance rotary inverters. Fully regulated output power is maintained with inputs from 21 to 36.4 VDC.

Solid-state inverter, model S1-2500B, is a TSO-approved unit. Fully regulated output power is maintained with inputs from 20 to 36 VDC and at least 100 vac under full load is maintained with inputs as low as 18 VDC. The S1-2500B can operate into equipment short circuits until the short clears or the circuit breaker opens.

The S1-3000A solid state inverter maintains fully regulated output with inputs from 20 to 36 VDC and at least 100 vac under full load with inputs as low as 18 VDC. This is sufficient voltage to operate gyroscopic flight instruments. This model can operate into equipment short circuits until the short clears or the circuit breaker opens.

The S1-1250A solid state inverter is a TSO-approved unit. Fully regulated output power is maintained with inputs from 24 to 30 VDC. At least 95 vac under full load is maintained for five minutes with inputs as low as 20 VDC.

The Eagle E.L.T. comes in several models:

The standard Eagle P/N 750065 has an RF test light for testing the E.L.T. without need of special equipment. It has a sealed battery pack. With an external antenna, the P/N 750071 can transmit for greater distances. With voice transmission, the P/N 750063 allows the aircraft microphone to be plugged into the ELT and voice transmitted to a rescue aircraft. The P/N 750071 has an external antenna and voice transmission feature.

The Dolphin transmitter can be water and manually activated, has a self-test light, is capable of being air dropped, and includes floats.

Emergency Locater Transmitter, Emergency Position Indicating Radio Beacon, Emergency Survival Transceiver

SPECIFICATIONS

Power output:	600 mw peak per channel
Frequency:	121.5 mHz and 243.0 mHz simultaneously crystal controlled
Battery:	Standard Eveready No. 560 pack—2 year service life (Alkaline)
Operating temperature:	150°F to −40°F (66°C to −40°C)
Transmitting range:	200 miles (320 km) (line of sight)
Activation:	Water-activated pressure switch and on-off test switch
Test circuit:	RF self-test light checks RF power output when switch is in test position
Transmitting life:	100 hrs.
Approvals:	FAA TSO'd to TSO C-91 Type S, Canadian approved RSS 147, Issue No. 2, Type W, meets ICAO requirements, FCC type accepted, part No. 87
Tether:	30 ft. (9.14 m.)
Capable of being air dropped:	With impact cone installed unit is capable of being dropped into water by aircraft flying at 500 ft. (152 meters) and 130 knots

The Whaler emergency position-indicating radio beacon is a float-free radio beacon.

SPECIFICATIONS

Power output:	75 mw peak per channel end of 48 hours
Frequency:	121.5 mHz and 243.0 mHz simultaneously crystal controlled
Battery:	Self-contained 7.5 volt pack (alkaline)
Battery service life:	2 years
Operating temperature:	+131° F to −4° F (+55° C to −20° C)
Transmitting range:	200 miles (320 km) (line of sight)
Activation:	Water activated pressure switch
Test circuit:	RF self test light checks RF power output when switch is in test position
Transmitting life:	100 hrs.
Approvals:	U.S. Coast Guard, Part 161.011-10, FCC Type Accepted under Part 83
Tether:	30 ft. (9.14 m)
Warranty:	2 years after original purchase by end user

Mako transmits a distress signal over a 200-mile radius for eight days while it floats in water.

SPECIFICATIONS

Operating frequencies:	121.5 mHz and 243.0 mHz.
Operating life:	48 hrs. min. (at −20°C −4°F) over 8 days while floating in water @ 75°F (@ 24°C)
Power output:	75mw min.—end life
Power source:	6.0V sealed battery pack, alkaline
Battery service life:	18 months (factory fresh batteries)
Transmitter duty cycle:	Continuous
Activation:	Manual on/off toggle switch
Transmitter indicator:	Light emitting diode (LED)
Antenna radiation:	Omnidirectional, telescopic, collapsible into case.
Range:	Up to 200 miles (322 km)—40,000 sq. miles (104,000 sq. km)

The EB-2B Survival ELT transmits distress signals over 40,000 square miles.

SPECIFICATIONS

Dimensions:	7-3/8" x 3-1/2" x 1-1/2"
Weight:	2.5 lbs.
Frequency:	*121.5 mHz and 243.0 mHz simultaneously
Frequency tolerance:	±.005%
Power output:	300 mw on each frequency
Modulation:	700 Hz downward sweep between 1600 Hz and 300 Hz at 3 sweeps per second
Range:	Swept tone—40,000 sq. miles Voice—line of visual sight
Operational life:	100+ hrs.
Batteries:	5 "C" size alkaline—1 year service life
Antenna:	Collapsible—Telescopic
Aluminum case:	Waterproof and shock resistant
Test feature:	R.F. energy illuminates lamp, which checks all circuitry to verify unit is operating properly
FCC type accepted:	Part 87
FAA TSO'd:	TSO'd C61A

*Other frequencies available for special applications

Martech also manufactures an air-deployable electronic Datum Marker Buoy with speeds up to 150 knots and altitude up to 500 feet.

Martech, Inc.
P.O. Box 1539
812 N.W. First Street
Fort Lauderdale, Florida 33302

MIDWEST PARACHUTE AVIATION PRODUCTS

Midwest Parachute is a certified parachute loft, licensed by the FAA, with approved loft ratings in parachute inspection, assembly, repair, certification, harness maintenance, modification, metal parts, and container overhaul as well as drop testing. A catalog from the company illustrates the different parachute models, plus goggles, luggage, canopies, and clothing.

Midwest Parachute Aviation Products
22799 Heslip Drive
Novi, Michigan 48050

TED NELSON COMPANY

The oxygen face masks provide optimum comfort and fit and improved safety during landing and takeoff by allowing radio transmission with "Hands on Controls." Sanitary conditions are enhanced through the use of the reservoir bag system.

Aircraft owners should choose the mask suited for their plane. For example, model E-5 is for aircraft with overhead oxygen outlets, such as the Cessna T-210, Bellanca, Mooney.

A new feature of the Nelson masks is that for the first time pilot and passengers can visually monitor the volume of oxygen flow in LPM as it flows to the face mask. The meters are equipped with a flow control needle valve which allows users to adjust the oxygen flow to correspond with the altitude the aircraft is flying at.

Ted Nelson Company
8638 Patterson Pass Road
Livermore, California 94550

Ted Nelson Company Oxygen Mask

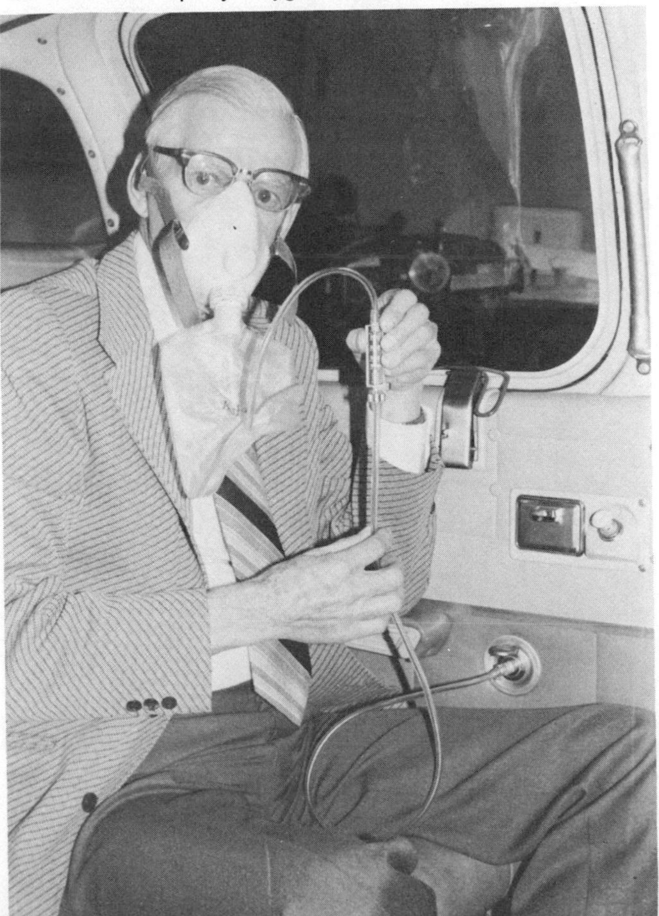

PILOTS SUPPLY COMPANY

Through its sister company, PDQ Computer Products, Pilots Supply offers several products to make a pilot's flight more pleasurable.

The PDQ Visual Airport Guide, made of heavy vinyl, is a two-second visual guide to the letdown area, runway, and traffic pattern.

The Flight Director includes space for writing a complete flight plan, flight progress record, ETA frequencies. Information on omniaids, radar assistance, scales, weather symbols, and emergencies can also be added. Made of heavy vinyl, the board can be erased after a flight.

The Mayday recording of the original tapes of the actual radio conversations between pilots and controllers during flight emergencies helps a pilot figure out how he would react in a flight emergency.

The company also sells sunglasses for pilots and an IFR hood.

Pilots Supply Company
Box 760 S
Asbury Park, New Jersey 07712

PURITAN-BENNETT AERO SYSTEMS CO.

Puritan-Bennett manufactures life support systems for both commercial and general aviation, including portable units, masks, emergency oxygen system control panels for the cockpit area of jet aircraft. In addition, the company also manufactures high pressure port fittings, line assemblies, charge adapters, leak detector fluids.

A few examples follow:

The Puritan-Bennett "Aero-Med" portable oxygen system dispenses therapeutic medical oxygen prescribed by physicians for their patient/passengers. The system delivers oxygen at flow rates up to eight liters per minute. The unit will also dispense prescription aerosol misting of oxygen when used in conjunction with disposable type humidifiers. The oxygen volume is controlled by a manually adjusted regulator, calibrated to provide a minimum of the selected flow at cabin altitudes from sea level to 8,000 feet. A fully charged cylinder, containing 625 liters of oxygen, will deliver a continuous flow for a maximum of four hours and 22 minutes at the rate of two liters per minute.

The Puritan-Bennett Sweep-On quick-donning flight crew oxygen mask has been upgraded through several of its key harness parts so that its life expectancy is longer.

The "Halo" quick-donning oxygen crew mask can be put on with one hand in two seconds, including over eyeglasses. A circular suspension ring provides for universal fit.

Write for the catalog detailing all prices:

Puritan-Bennett Aero Systems Co.
111 Penn
El Segundo, California 90245

PDQ Visual Airport Guide

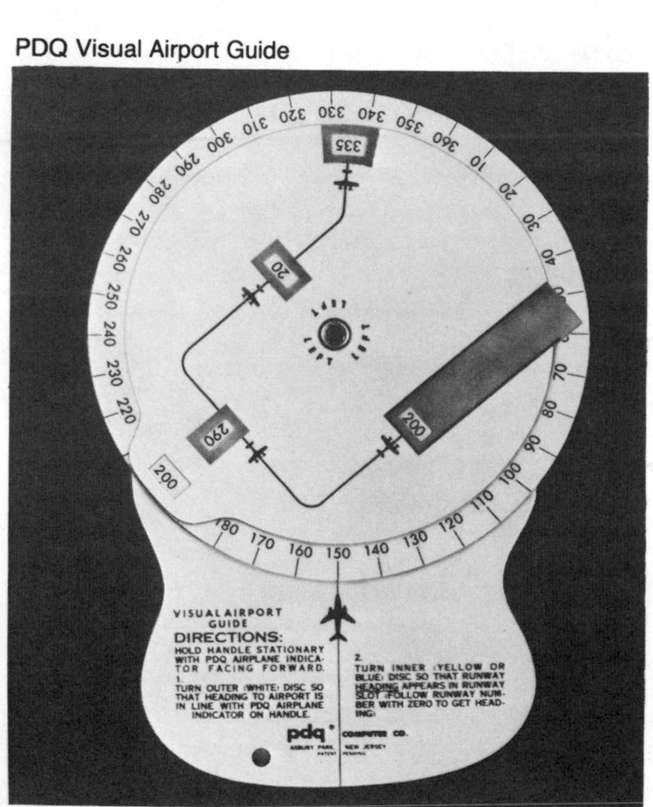

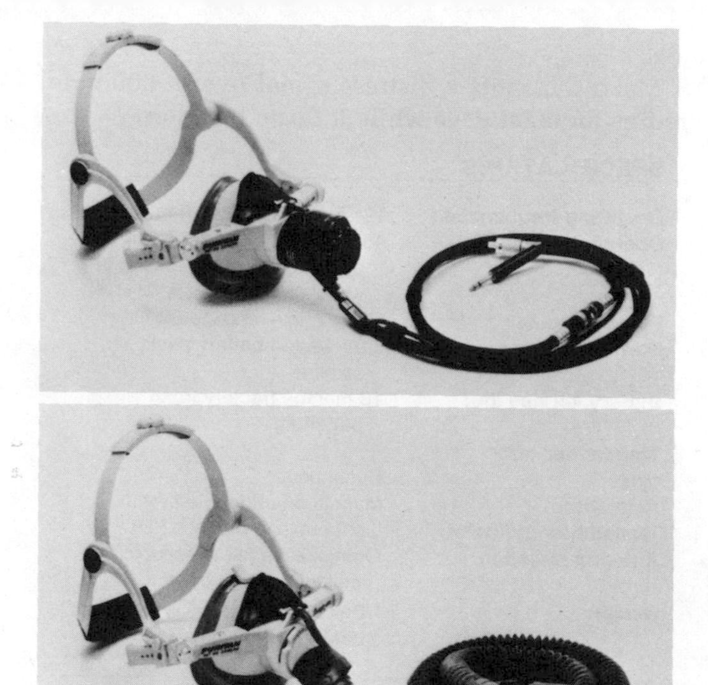

Puritan-Bennett Crew Mask

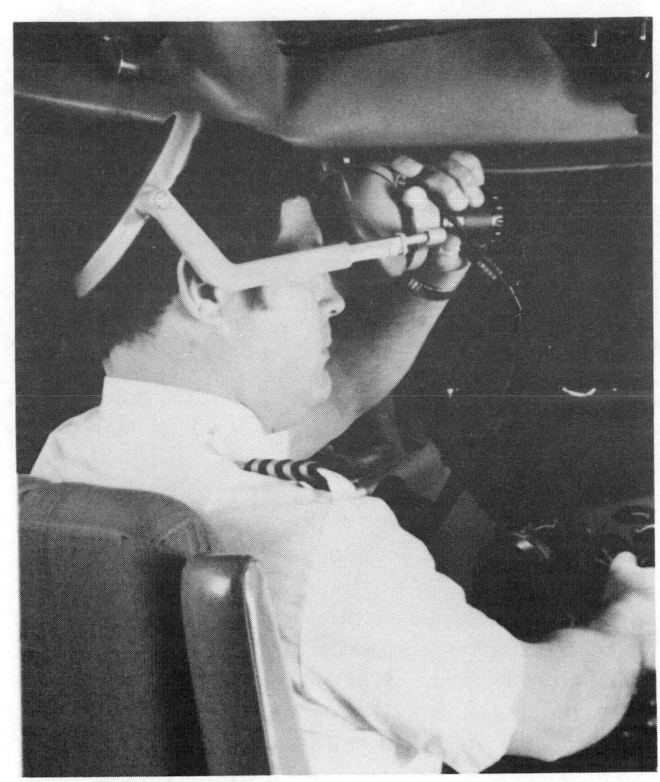

"Halo" Quick Donning Mask from Puritan Bennett

SABRE RESEARCH

Sabre Research's Flite Minder Airport Data table provides a place where pilots can jot down important information in the exact sequence in which it is transmitted from the airport tower. In addition, pilots must consider weather conditions, which can, and often do change between takeoff and landing. With the Flite Minder Weather Data table, the pilot can copy all significant weather data. Finally, there is a Flite Minder Winds Aloft where a pilot can note the direction and velocity of the winds.

The Flite Minder forms can fit conveniently on kneeboard or in the control yoke clip. A pad of 50 forms packaged with instructions measures 5½-by-8½ inches.

Sabre Research
P.O. Box 510
Jamaica, New York 11431

SCOTT AVIATION

Scott manufactures oxygen systems for safe airline travel. These include portable oxygen systems, oxygen mask dispensing units, oxygen breathing equipment, Aviox single-pak and duo-pak portable oxygen breathing units, emergency escape breathing devices, rubber oral-nasal masks, airline crew oxygen masks.

In addition, Scott manufactures tailwheel assemblies and accessories, instruments and gauges, first aid kits, fire extinguishers, brake master cylinders, parking brake valves, sanitary disposable relief containers and cylinder and valve assemblies.

Several examples of the oxygen systems follow.

Executive Mark I supplies constant flow oxygen for one or two persons up to altitudes of 16,500 feet. The unit features a finger-operated top-mounted on/off valve.

The Executive Mark III provides automatic altitude compensated oxygen at a constant flow for one to four persons. It operates up to 19,000 feet or to 27,000 feet with optional accessories. Features include finger-operated front-mounted on/off valve, soft, rubberlike oral-nasal mask with flow indicator, and seat-mounted console case.

Flite Minder Airport Data Table

FLITE MINDER ™ No. ______

Date ____________ Aircraft No. ____________

AIRPORT DATA

	Origin	Destination
Active Runway		
Wind Direction (degrees)		
Wind Velocity (knots)		
Altimeter Setting (in. Hg)		
Time (Z)		

WEATHER DATA

Location	Time (Z)	Ceiling (ft)	Visibility (s. mi.)	Clouds	
				Base (ft)	Top (ft)
Origin					
Enroute					
Destination					
General Weather					

WINDS ALOFT

Altitude (feet)	Origin		En Route		Destination	
	Dir. (deg)	Vel. (kt)	Dir. (deg)	Vel. (kt)	Dir. (deg)	Vel. (kt)
3,000						
6,000						
9,000						
12,000						
Time (Z)						

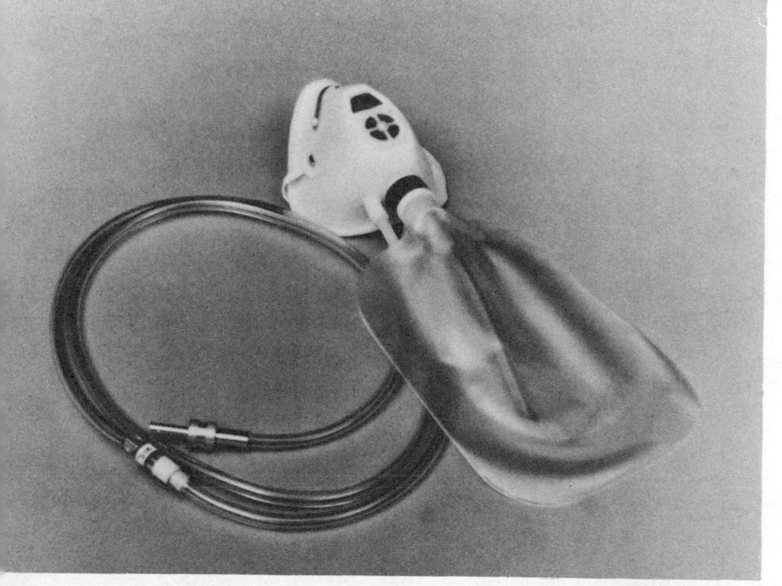

Scott Aviation Skymask

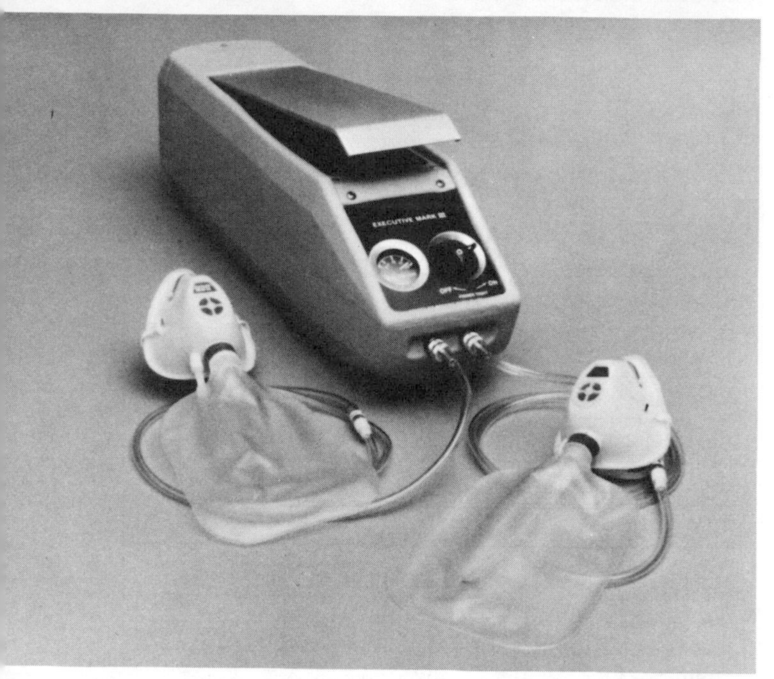

Executive Mark III Mask

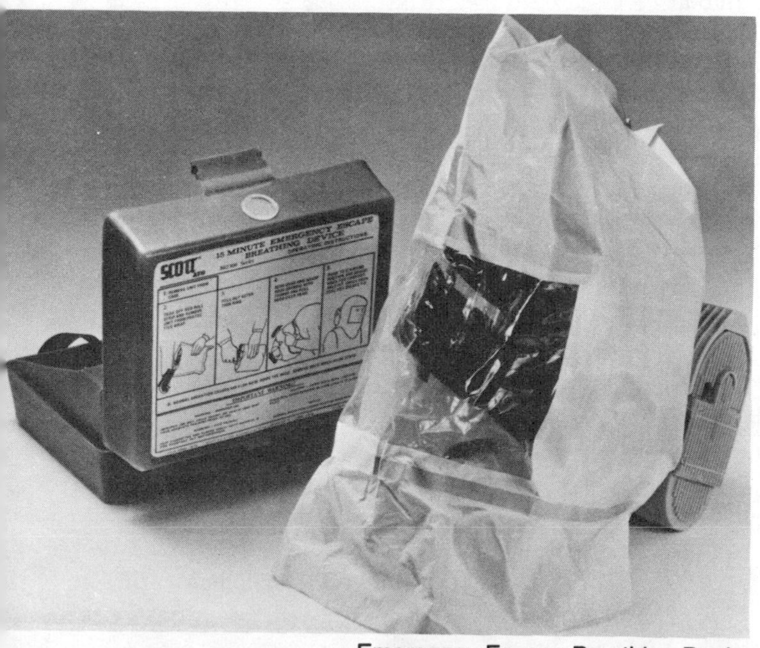

Emergency Escape Breathing Device

The Sky Mask is made of soft rubberlike material with an oral-nasal facepiece with built-in aluminum nose clip to provide a custom fit. Available with a microphone kit that can be installed at the factory or by the user, the Sky Mask includes replaceable items—rebreather bag, filter, elastic band assembly, and filter retainer.

The Scott Folding Quick-Don Assembly comes with a choice of face masks. The 358-1002 model has narrow cheek flaps. The 358-1030 model has wide cheek flaps with an extra wide face seal flange and is commonly used with a mask mounted regulator.

The Scott Vent-Valve system provides an option to reduce smoke or noxious fumes that may clog the smoke goggle cavity during certain emergency situations. Operated by a manual push-pull control knob located on the nose piece of the mask, the vent-valve diverts a small amount of the oxygen flow from the mask into the goggle area, raising the pressure and purging smoke or fumes to the outside.

The Emergency Escape Breathing Device is designed for use by aircrew members for short durations in an oxygen-deficient or irrespirable atmosphere. It is an independent 15-minute, self-contained, completely disposable breathing apparatus designed to provide respiratory and eye protection to the user in oxygen-deficient, smoke-laden, or other toxic atmosphere. The system consists of four major components: solid-state oxygen supply source; a chemical carbon dioxide and water vapor "scrubber"; a loose-fitting hood with neck seal to enclose the head and provide respirable environment for the user; and a venturi "pumping" arrangement, powered by the oxygen generator, that provides make-up oxygen and recirculates the gas within the system loop consisting of the scrubber and the hood.

The Aviox Single-Pak and Duo-Pak units utilize solid state oxygen generators, providing immediate oxygen availability without high pressure oxygen storage. Features include actuator lock/unlock button; safety check valve which allows removal and insertion of one generator while the second is flowing, without loss of oxygen; aircraft relief valve with standard valve head; comfortable mask, which can be easily detached; dual purpose carrying strap.

The Single-Pak provides 4 LPM (STPD) flow for approximately 20 minutes. Duo-Pak provides 40 minutes when generators are initiated separately or 8 LPM flow for 20 minutes when both generators are initiated simultaneously.

Scott Aviation
225 Erie Street
Lancaster, New York 14086

L.A. SCREW PRODUCTS

The Power Probe flashlight, model PPB-2000, gives aircraft pilots precise light beams. The Smoke Cutter, model SCD-2000-L, provides a reliable optical system under extreme fire, water, and smoke conditions. Both models use D-cell batteries. A lanyard, made from 500-pound braid and available with a spare bulb, is an optional feature. The Smoke Cutter, model SCD-2000-L, provides reliable help in fire, water, and smoke conditions, concentrating light into a narrow smoke-penetrating beam. The Code Four D-cell flashlight is waterproof to a 100-foot-depth, is explosion-proof certified and has a recessed "fail-safe" switch. In addition, battery barrels can be added or subtracted to form the size flashlight needed. A spare bulb is protected by a steel guard in the tail cap.

L.A. Screw Products
Police Equipment Division
8401 Loch Lomond Drive
Pico Rivera, California 90660

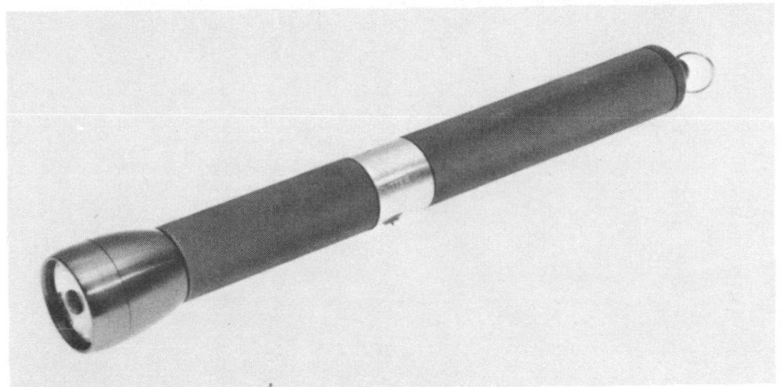

L.A. Screw Products Power Probe Flashlight

TANIS AIRCRAFT SERVICES

Tanis has developed a heating system that allows an aircraft owner to plug in the airplane to an electrical outlet so that the plane is always ready to go. The four cylinder standard unit draws only 250 watts of power, about the same power consumption as a light bulb. Heat is transmitted by conduction.

The unit is not intended to rapidly preheat an engine, but takes five to six hours. All the equipment an owner needs is a drop cord and a common blanket.

The following engines can be fitted:

- All Lycoming: O-320, O-360, IO-320, and IO-360, O-540 and IO-540 series, TIO-541 and TIGO-541
- Continental: O-470, IO-470, and IO-520, IO-360 and TSIO-360, E-185 and E-225 (other engines in development)

Tanis also sells engine covers to make the preheater system work even better. The covers retain engine heat for three or four hours after shutdown without preheating. Insulated with dacron fiber, the covers weigh three to four pounds each. They are closed with velcro and are fitted to each aircraft model.

Tanis Heating System

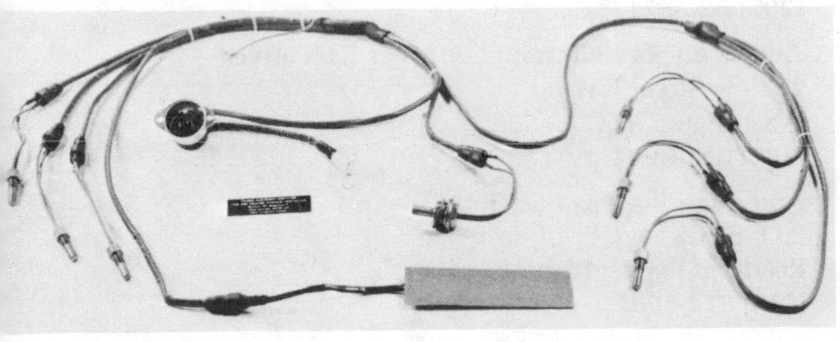

Tanis Aircraft Services
P.O. Box 117
Glenwood, Minnesota 56334

WILBUR INDUSTRIES

Only 18 pounds, the preheater engineered by Wilbur Industries is low on fuel consumption (45 minutes of preheat per cylinder at 325 degrees). Additional benefits include a 100CFM blower with low current drain, gas pressure gauge, outlet temperature gauge, which indicates temperatures of outlet air to airplane, and easy, fast operation.

Wilbur Industries
Southwest Harbor
Maine 04679

Wilbur Industries Preheater

5 / Aeronautical Associations and Publications

ASSOCIATIONS

There are hundreds of aeronautical associations a pilot can join, many of which are organized for a specific audience such as the Airborne Law Enforcement Association, American Ultralight Association, and the Christian Pilots Association. Following is an up-to-date list that includes membership fees and publications.

Academy of Model Aeronautics
815 15th Street NW
Washington, D.C. 20005
(202) 347-2751

Airborne Law Enforcement Association
2639 Maplewood Drive
Columbus, Ohio 43229
(614) 891-2782

Aircraft Electronics Association
P.O. Box 1981
Kansas City, Missouri 64055
(816) 373-6565

Aircraft Owners and Pilots Association (AOPA)
Air Rights Building
7315 Wisconsin Avenue
Bethesda, Maryland 20014
(301) 654-0500

Publishes the monthly magazine, *The Pilot*. Also offers domestic and international flight planning service, aviation research service, aircraft title search and insurance, flight training clinics, group vacation tours, newsletter. Membership: $29 per year.

Air Traffic Control Association
Suite 409
525 School Street SW
Washington, D.C. 20024
(202) 522-5717

American Association of Airport Executives
2029 K Street NW
Washington, D.C. 20006
(202) 331-8994

American Bonanza Society
P.O. Box 3749
Reading, Pennsylvania
(215) 372-6967

American Helicopter Society
Suite 103
1325 18th Street NW
Washington, D.C. 20036
(202) 659-9524

Publishes bimonthly news magazine, *Vertiflite*, and also a technical quarterly. Membership: $30 for first year, $25 for renewal of membership.

American Navion Society
P.O. Box 1175
Municipal Airport
Banning, California 92220

American Ultralight Association
P.O. Box 2287
Ogden, Utah 84404

Publishes quarterly newsletter with safety information and regulations.

Antique Airplane Association
Route 2
Box 172
Ottumwa, Iowa 52501
(515) 938-2773

Publishes quarterly magazine, *Antique Airplane Digest*. Membership: $30 per year.

Aviation Distributors and Manufacturers Association
1900 Arch Street
Philadelphia, Pennsylvania 19103
(215) 564-3484

Aviation Executives Club
500 Deer Run
Miami Springs, Florida 33166

Aviation Maintenance Foundation
P.O. Box 739
Basin, Wyoming 82410
(307) 568-2466

Aviation/Space Writers Association
Cliffwood Road
Chester, New Jersey 07930
(201) 879-5667

Publishes newsletter and manual. Membership: $40 plus $10 initiation.

Balloon Federation of America
Suite 430
821 15th Street NW
Washington, D.C. 20005
(202) 737-0897

CESSNA International 120-140 Association
Box 92
Richardson, Texas 75080

Publishes monthly newsletter. Membership: $10 per year.

China, Burma, India Hump Pilots Association
c/o Jan Thies, Executive Secretary
917 Pine Boulevard
Popular Bluff, Missouri 63901
(314) 785-2420

Christian Pilots Association
Box 5157
Pasadena, California 91107
(213) 797-9515

$25 dues per year.

Civil Air Patrol, Headquarters
Attention DO
Maxwell Air Force Base, Alabama 36112
(205) 293-5198

Civil Aviation Medical Association
801 Green Bay Road
Lake Bluff, Illinois 60044
(312) 234-6330

Confederate Air Force
Rebel Field
P.O. Box CAF
Harlingen, Texas 78550
(512) 425-1057

Early Birds of Aviation
c/o Paul Gerber, president
310 North Jackson Street
Arlington, Virginia 22201
(703) 524-5198

Ercoupe Owners Club
3557 Roxboro Road
P.O. Box 15058
Durham, North Carolina 27704
(919) 477-2193

Experimental Aircraft Association
P.O. Box 229
Hales Corners, Wisconsin 53130

Publishes monthly magazine, *Sport Aviation*. Has special interest groups: antique/classic division, international aerobatic club, Warbirds of America. Membership: $25 per year.

Flying Chiropractors Association
c/o Frederick J. Miscoe
215 Belmont Street
Johnston, Pennsylvania 15904
(814) 266-3314

Flying Dentists Association
5820 Wilshire Boulevard
Los Angeles, California 90036
(213) 937-5514

Flying Optometrist Association of America
c/o Dr. Howard Flippin
311 North Spruce
Searcy, Arkansas 72143
(501) 268-3577

Flying Physicians Association
801 Green Bay Road
Lake Bluff, Illinois 60044
(312) 234-6330

Flying Psychologists
190 North Oakland Avenue
Pasadena, California 91101
(213) 795-5144

For $3, a psychologist can become a member and receive the group's newsletter.

General Aviation Manufacturers Association
1025 Connecticut Avenue NW
Suite 517
Washington, D.C. 20036
(202) 296-6540

Helicopter Association of America
1110 Vermont Avenue N.W.
Suite 430
Washington, D.C. 20005
(202) 466-2420

International Aerobatic Club
P.O. Box 229
Hales Corners, Wisconsin 53130
(414) 425-4860

International Comanche Society
M.E. Tipton, M.D.
600 East 4th Street
Bellwood, Pennsylvania 16617
(814) 742-8446

International Flying Farmers
Mid Continent Airport
P.O. Box 9124
Wichita, Kansas 67227
(316) 943-4234

Lawyer-Pilots Bar Association
815 12th Street
Lawrenceville, Illinois 62439
(618) 943-2338

National Aeronautics Association
Suite 430
821 15th Street, NW
Washington, D.C. 20005
(202) 347-2808

National Agricultural Aviation Association
Suite 459
National Press Building
Washington, D.C. 20045
(202) 638-0542

National Association of Flight Instructors
Ohio State University Airport
Box 20204
Columbus, Ohio 43220
(614) 459-0204

National Association of State Aviation Officials
Suite 400
1300 G Street, NW
Washington, D.C. 20005
(202) 783-0588

National Business Aircraft Association
One Farragut Square South
Washington, D.C. 20006
(202) 783-9000

Ninety-Nines, Inc.
Will Rogers World Airport
Oklahoma City, Oklahoma 73159
(405) 685-7969
and
107 Wilmington Drive
Melville, N.Y.
(516) 643-2692

Organized Flying Adjusters
c/o John W. Axe, president
P.O. Box 14985
Austin, Texas 78761

OX5 Aviation Pioneers
605 Allegheny Building
Pittsburgh, Pennsylvania 15219
(412) 281-6477

Pilots International Association, Inc.
400 South County Road 18
Minneapolis, Minnesota 55426

Membership fee of $15 per year includes subscription to monthly magazine, *Plane & Pilot*. Membership also includes quarterly newsletter *Flight Line*, low-cost aircraft title search service, free film, library, member chart service, aircraft finance plan, travel service.

Popular Rotorcraft Association
P.O. Box 570
Stanton, California 90680

Publishes six magazines a year, *Popular Rotorcraft Flying*, and organizes an annual fly-in. Also teaches its members safe building and flying habits. Membership: $18 per year, includes subscription to the magazine.

Professional Aviation Maintenance Association
P.O. Box 248
St. Ann, Missouri 63074
(314) 426-7060

Soaring Society of America, Inc.
Box 66071
Los Angeles, California 90066
(213) 390-4447

Publishes monthly magazine, *Soaring*, plus directory of soaring sites and organizations and safety and training programs. Also sells hats, jackets, T-shirts, calculators and art. Membership: $28 per year.

Society of Experimental Test Pilots
P.O. Box 986
Lancaster, California 93534

Taildragger Pilots Association
P.O. Box 161079
Memphis, Tennessee 38118
(901) 755-8612

Membership fee of $15 entitles taildragger to participate in taildragger flying clubs; discounts up to 50 percent on aircraft parts, magazines, newsletter called *Taildragger Tales*, pilot supplies, and insurance; membership card honored for rental car discounts. Also sells such items as T-shirts, glassware.

United States Hang Gliding Association
P.O. Box 66306
Los Angeles, California 90066

$25 membership fee.

United States Parachute Association
Suite 444
806 15th Street NW
Washington, D.C. 20005
(202) 347-5773

Publishes monthly magazine, *Parachutist*. Also offers jump pilot's manual, insurance, patches, buckles, bumper stickers, T-shirts, film and posters. Membership: $26.50 per year.

United States Seaplane Pilots Association
P.O. Box 3009
Washington, D.C. 20014
(301) 654-0500

Whirly-Girls
Suite 700
1725 Desales Street NW
Washington, D.C. 20036
(202) 347-2315

MAGAZINES

A/C Flyer
444 Brickell Avenue
Miami, Florida 33131

Sent to owners of registered airplanes for free. Otherwise, 12 issues cost $15 or $1.75 for a single issue.

Air Power
Sentry Books
10718 White Oak Avenue
Granada Hills, California 91344

Published six times a year (alternating with Wings magazine). Subscription: $11. Single issue: $1.95. Covers subjects of aviation history.

Air Power Museum Bulletin
Route 2
Box 172
Ottumwa, Iowa 52501

Published four times a year for members of Antique Airplane Association. If not member, $7.50 subscription.

Air Progress
Challenge Publications
7950 Deering Avenue
Canoga Park, California 91304

Monthly subscription is $11.98; single copies, $1.50.

Aircraft Directory
Werner & Werner Corporation
606 Wilshire Boulevard
Santa Monica, California 90401

Written by the editors of *Plane & Pilot* and published once a year with up-to-date information on all general aviation airplanes. Cost: $3.95.

Airman's Information Manual
Werner & Werner
606 Wilshire Boulevard
Santa Monica, California 90401

Published once a year for $5.50, plus $2 for postage and handling. Includes FAA regulations.

Airport Services Management
731 Hennepin Avenue
Minneapolis, Minnesota 55403

Published 12 times a year, $18 subscription.

Airports International
IPC Business Press
U.S. Correspondent
205 East 42nd Street
New York, New York 10017

Published eight times a year for $58.10 (via air mail, from England). International reporting of all phases of civil air transport.

Antique Airplane Digest
Route 2
Box 172
Ottumwa, Iowa 52501

Members of Antique Airplane Association receive this quarterly magazine free of charge.

Aviation
P.O. Box 186
Brookfield, Connecticut 06804

Published 12 times annually, $12.95 subscription. Includes extensive list of airplanes for sale.

Aviation Monthly
757 Third Avenue
New York, New York 10017

Published 12 times annually, $12 subscription.

Business and Commercial Aviation
One Park Avenue
New York, New York 10016

Published 12 times annually, $24 subscription, but individual copies can be purchased. Good source for buying and selling an airplane.

Flight International
IPC Business Press
205 East 42nd Street
New York, New York 10017

Weekly publication costing $98.80 for a subscription.

Flying
One Park Avenue
New York, New York 10016

Published 12 times a year: $12.97 for subscription, $1.50 for single copies.

Flying, Annual & Buyers' Guide
One Park Avenue
New York, New York 10016

Published once a year, $2.50.

Hang Gliding
P.O. Box 66306
Los Angeles, California 90066

Free if member of U.S. Hang Gliding Association. If not, $18 for 12 issues.

Home Built Aircraft
Werner & Werner Corporation
606 Wilshire Boulevard
Santa Monica, California 90401

Published 12 times a year. Special offer of 12 copies for $11.70. Individual copies for $1.95.

Parachutist
806 15th Street
N.W., Suite 444
Washington, D.C. 20005

Publication of U.S. Parachute Association, 12 times a year for $28.50 or $2 for a single issue. Members of Parachute Association receive copies for free.

The Pilot
7315 Wisconsin Avenue
Washington, D.C. 20014

Monthly magazine for members of Aircraft Owners & Pilots Association.

Pilot News
Suburban Pilot Inc.
5320 N. Jackson Avenue
Kansas City, Missouri 64119

Comes out monthly: $9 for one-year subscription, $15 for two-year subscription, $18 for three-year subscription. No individual copies sold.

Same group also publishes *Aircraft Owners Publication* monthly in three editions for Iowa and Nebraska, Kansas, Missouri, and Southern Illinois, and Oklahoma and Northern Texas.

A third publication is *Kansas City Aviation*, distributed free monthly to the 5,000 top business people in Kansas City. It is available to others for $13.50 for one year, $24 for two years, and $33 for three years.

Pilot Reports
Challenge Publications
7950 Deering Avenue
Canoga Park, California 91304

Quarterly magazine at $9 for subscription or $3 for single copy.

Plane & Pilot
Werner & Werner
606 Wilshire Boulevard
Santa Monica, California 90401

Published 12 times a year, special offer of 12 issues for $7. Individual copies: $1.50. Members of Pilots International Association receive magazine for free.

Private Pilot
8322 Beverly Boulevard
Los Angeles, California 90048

Published 12 times annually for $13.97. Single copies: $1.50.

Rotor & Wing International
News Plaza
Box 1790
Peoria, Illinois 61656

Published 12 times a year, $18 subscription. Magazine for helicopter owners and enthusiasts.

Soaring
Box 66071
Los Angeles, California 90066

Monthly magazine for members of Soaring Society of America.

Sport Aviation
P.O. Box 229
Hales Corners, Wisconsin 53130

Monthly magazine for members of Experimental Aircraft Association.

Vertiflite
1325 18th Street N.W.
Washington, D.C. 20036

Bi-monthly news magazine for members of American Helicopter Society.

The Weekly of Business Aviation
1156 15th Street N.W.
Washington, D.C. 20005

$165 for a subscription, with weekly publication. Single copies are $3. Covers news of new aviation products, services, employment, maintenance tips and up-to-date information from the CAB.

Wings
Sentry Books
10718 White Oak Avenue
Granada Hills, California 91344

Published six times a year (alternating with *Air Power* magazine). Subscription: $11. Single issue: $1.95. Covers subjects of aviation history.

CATALOGS AND BOOKS

Browse through the following catalogs and you will find almost every imaginable item you'll ever need or want for your airplane or home: parts and tools for planes, clothing with an aviation motif, books about aviation, dishes, glasses, and much more. Almost all of the catalogs are available free of charge.

Aero Publishers
329 West Aviation Road
Fallbrook, California 92028
(714) 728-8456
Free catalog listing more than 200 books on aviation-related subjects.

Aircraft Components Flyer
700 North Shore Drive
Benton Harbor, Michigan 49022
1-800-253-0800
Free catalog/magazine advertising instruments, books, glasses, clothing, kneeboards. Published six times annually.

Aircraft Spruce and Specialty Company
Box 424
Fullerton, California 92632
(714) 870-7551
$4 catalog, the cost of which is applied toward purchases greater than $25. New catalog comes out each July. Items listed in the more than 200-page catalog include wood products, hardware, airframe parts, engine parts, tools, pilot supplies, plus more—all for the person who wants to build his or her own airplane.

The Airplane Shop, Inc.
125 Passaic Avenue
Fairfield, New Jersey 07006
(201) 575-9621 or 736-9092
Free catalog lists hundreds of items for the airplane owner or aviation enthusiast from stainless steel bolts to original Army Air Corps wings and other memorabilia.

Airtex
259 Lower Morrisville Road
Fallsington, Pennsylvania 19054
(215) 295-4115
Free catalog showing Airtex's custom-designed interiors for airplanes, plus wind socks, protective covers, aircraft finishes.

All Aircraft Parts
16673 Roscoe Boulevard
Sepulveda, California 91343
(213) 894-9115
$2 catalog lists hundreds of items for an aircraft owner—fittings, bolts, avionics equipment, adhesives, hardware, and accessories like kneeboards, charts, flight cases.

Aviation Book Catalog
1640 Victory Boulevard
Glendale, California 91201
(213) 240-1771
Free catalog lists 1,100 new aeronautical books, plus pilot supplies and flight aids.

Aviation Buyers Directory
1 Bank Street
Stamford, Connecticut 06901
(203) 325-2647
Quarterly catalog, provides a directory of parts and manufacturers, plus aircraft services. Distributed to aviation industry purchasing executives.

Aviation Distributors and Manufacturers Association
1900 Arch Street
Philadelphia, Pennsylvania 19103
(215) 564-3484
$15 Membership Directory of Aviation Distributors and Manufacturers Association (ADMA).

Airman's Information Manual from
Aviation Maintenance Publishers

Private Pilot Airplane from
Aviation Maintenance Publishers

Basic Helicopter Handbook from
Aviation Maintenance Publishers

Answers and Explanations for Private Pilot Airplane from
Aviation Maintenance Publishers

EA-AC 61-32CG

ANSWERS
AND
EXPLANATIONS
FOR
PRIVATE PILOT
AIRPLANE
Written Test Guide

AVIATION MAINTENANCE PUBLISHERS, INC.
P.O. Box Basin, WY 82410
(307) 568-2413

Aviation Maintenance Publishers
211 South Fourth Street
Basin, Wyoming 82410
(307) 568-2413 or toll free 1-800-443-9250

Publishes a technical book for free, listing educational training materials.

Private Pilot Airplane from Aviation Maintenance Publishers

Aviation Products
114 Bryant
Ojai, California 93023
(805) 646-6042

Free catalog of parts and tools for both new and used aircraft.

The Aviation Telephone Directory
9355 Chapman Avenue, Room 202
Garden Grove, California 92641
(714) 539-8931

Free directory listing airports, aerodomes, and aviation-related firms.

Century Instrument Corporation
440 Southeast Boulevard
Wichita, Kansas 57210
(316) 683-7571

Free price list of new and rebuilt instruments.

Understanding the Federal Air Regulations from Aviation Maintenance Publishers

The Cockpit, Division of Avirex
468 Park Avenue South
New York, New York 10016
(212) 686-7707

Free catalog of clothing, insignias, jewelry, binoculars, clocks, headsets, knives, survival lights, art posters.

Cooper Aviation Supply Company
2149 East Pratt Boulevard
Elk Grove Village, Illinois 60007
(312) 364-2600

Free catalog of aircraft parts for the do-it-yourself builder: primers, tape, wall panel sets, safety belts, tubing, hose fittings, electronics, seat cover sets.

Everybody's Bookshop
317 W. 6th Street
Los Angeles, California 90014

Out-of-print books and magazines and used originals in good condition, sold on a first-come, first-served basis. On flying: *Aviation Week & Space Technology, Aero Digest, Air Classics, Air News, Air Progress, Air Trails, American Modeller, American Aircraft, Flying, Flying Aces, Flying Models, Model Airplane News, Model Builder, Plane & Pilot, Private Pilot, R.C.M.*, and more.

Flight Apparel Industries
P.O. Box 166
Hammonton Airport
Hammonton, New Jersey 08037
(609) 561-9200
Free catalog of uniforms for crew and passengers and company blazers.

Flight Log
Box 39675
Los Angeles, California 90039
(213) 633-7561
Bimonthly magazine costing $18 a year details places to travel in an airplane, luggage, tools, wax, where to go for fly-ins, maintenance tips.

Fudpucker
Dept. PV 1279
Box 67
Hayden Lake, Idaho 83835
(208) 772-2211
Free catalog of clothing, flight bags, glassware, art with aviation motif.

Goldstar Enterprises
P.O. Box 363
Williamsville, New York 14221
Offers two publications: *Taming the Taildragger*,a flight manual for classic tailwheel aircraft; and *Getting a Flying Start in Aviation Business,* which describes part-time aviation businesses. *Taming the Taildragger*: $5.95. *Getting a Flying Start in Aviation Business*: $8.95.

Hidalgo's
Hidalgo Supply Company
Dept. C-10
P.O. Box 35339
Houston, Texas 77035
(713) 729-6940
Free catalog of such items as knives, sunglasses, binoculars, headsets, scanners, space heaters.

Pilot Logbook from Flight Log

Iowa State University Press
South State Avenue
Ames, Iowa 50010
Sales Department has a list of aviation publications offered, including: *Buying and Owning Your Own Airplane* by James E. Ellis, $12.95; *The Student Pilot's Flight Manual* by W.K. Kershner, $12.95; *The Advanced Pilot's Flight Manual* by W.K. Kershner, $13.95; *Fly the Wing* by Jim Webb, $14.50; *The Instrument Flight Manual* by W.K. Kershner, $12.50; *ATP Airline Transport Pilot* by K.T. Boyd, $13.50; *The Flight Instructor's Manual* by W.K. Kershner, $14.95; *West to the Sunrise* by Grace Harris, $12.50; *Pilot Logbook* by W.K. Kershner, $4.25

J & M Aircraft Supply
1037 Hawn Avenue
P.O. Box 7586
Shreveport, Louisiana 71107
(318) 222-5749
Free catalog of materials and parts for do-it-yourself builders: avionics equipment, logbooks, sunglasses, flight cases, maps, clocks, books, portable oxygen systems, parts, polish, paint, accessories.

The Instrument Flight Manual

Buying and Owning Your Own Airplane

Kaybro Sales Company
P.O. Box 24916
Los Angeles, California 90024
(213) 476-2463
$1 catalog (except free for AOPA members) listing jewelry, prints, gifts, mobiles, glasses, scale replicas and other aviation-oriented material.

Manual List
Essco
Akron Municipal Airport
Akron, Ohio 44306
(216) 733-6241
Free catalog of manuals to buy on airframes, engines, and other such subjects.

National Flightshops
St. Petersburg-Clearwater Airport
Clearwater, Florida 33520
(813) 531-3545
Free flyer listing a sampling of this group's aeronautical potpourri: neckties, T-shirts, wall art, books.

The Student Pilot's Flight Manual

Fly The Wing

The Advanded Pilot's Flight Manual

Neckties from NATIONAL FLIGHTSHOPS

Pan American Navigation Service
16934 Saticoy Street
Van Nuys, California 91409
(213) 345-2744
Free catalog listing flashlights, belt buckles, maps, log books, computers and other aviation equipment.

Para-Gear
3839 W. Oakton Street
Skokie, Illinois 60076
(312) 679-5905
$1 catalog listing equipment for parachutists, including parachutes, clothing, hardware, pack opening bands, goggles, instruments, knives, telemeters, movies about parachuting.

Book from NATIONAL FLIGHTSHOPS

Book from NATIONAL FLIGHTSHOPS

Pittsburgh Institute of Aeronautics
Allegheny County Airport
P.O. Box 10897
Pittsburgh, Pennsylvania 15236
Aircraft Mechanic's Specifications Handbook for $7.95, compiled by an airline mechanic for use by mechanics. Group publishes other aviation-related books.

The Rand McNally Encyclopedia of Military Aircraft from Rand McNally & Co.

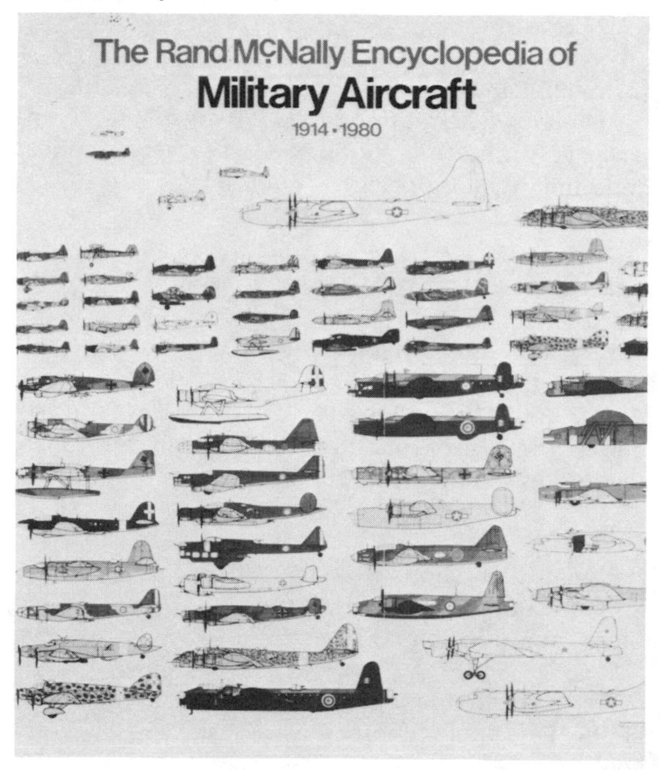

World Aircraft from Rand McNally & Co.

Rand McNally & Co.
P.O. Box 7600
Chicago, Illinois 60680
Offers a series of aviation guides as well as an Encyclopedia of Military Aircraft, 1914-1980. Send for free descriptive brochure.

Schweizer Aircraft Company
P.O. Box 147
Elmira, New York 14902
Schweizer offers an extensive list of soaring books for beginners through competitive levels. Schweizer also has a soaring school which runs from the beginning of May through September and operates at Chemung County Airport, north of Elmira. Write for their catalog of available books.

Skybuys
P.O. Box 4111
San Clemente, California 92672
Skybuys offers several books: *Primary Aerobatic Flight Training* by Lt. Col. Art Medore, USAFR, $8.75 plus postage and handling; *Airports of Baja California and Airports of Mexico and Centro America*, $24.95 plus postage and handling, includes information on 426 airports, charts, IFR plates; *Those Incomparable Bonanzas* by Larry A. Ball, $16.95 plus postage and handling, history of Beechcraft Bonanza; *The Aviation Art of Keith Ferris*, $.95, includes 40 color plates in 9-by-11-inch volume. Skybuys also offers back issues of *Aero* and *Private Pilot* magazines.

Split-S Aviation
1050-K Pioneer Way
El Cajon, California 92020
(714) 440-0894
Free catalog of goggles, flight jackets, patches, microphones, jewelry, lamps, buckles, glassware.

Sporty's Pilot Shop
Clermont County Airport
Batavia, Ohio 45103
(513) 732-2411
Free catalog comes out every three or four months, with such listings as books, glasses, training hoods, portable oxygen systems, avionics equipment, jewelry.

Taxlogs Unlimited
20 Galli Drive
Ignacio, California 94947
(415) 883-7768
Leaflets on products: pilot taxlogs, aviation fuel, insurance guides, books.

Tronair
South 1740 Eber Road
Holland, Ohio 43528
(419) 866-6301
Catalog of aircraft ground support equipment for jets, turboprops, helicopters, plus towbars, hydraulic jacks, power units, service units, tractors.

Book from SKYBUYS

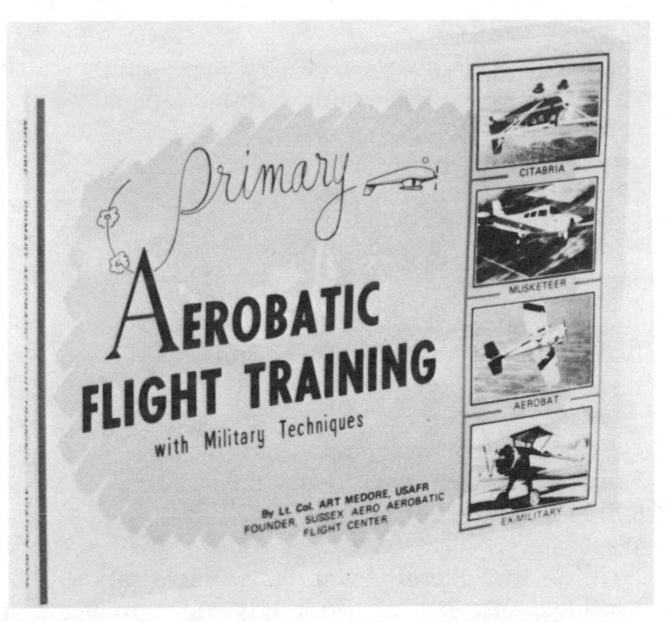

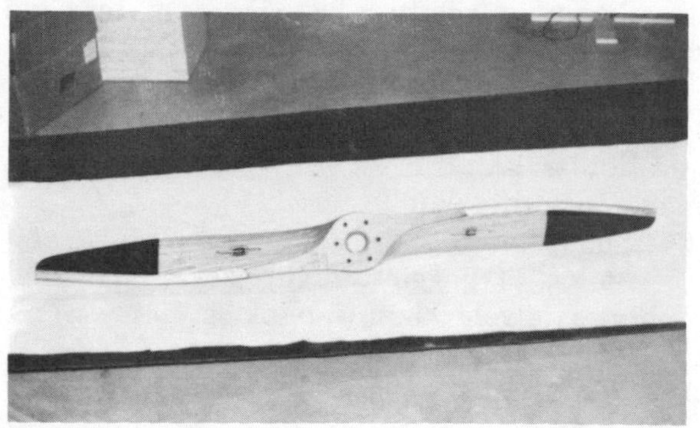

Wood Prop from UNIVAIR AIRCRAFT CORPORATION

Fuselage Cowl from UNIVAIR AIRCRAFT CORPORATION

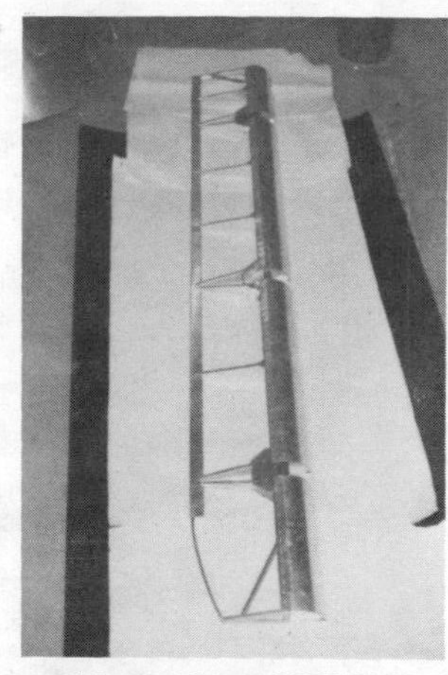

Aileron from UNIVAIR AIRCRAFT CORPORATION

Wing Prop from UNIVAIR AIRCRAFT CORPORATION

"UNIVAIR GENERAL PARTS CATALOGUE"

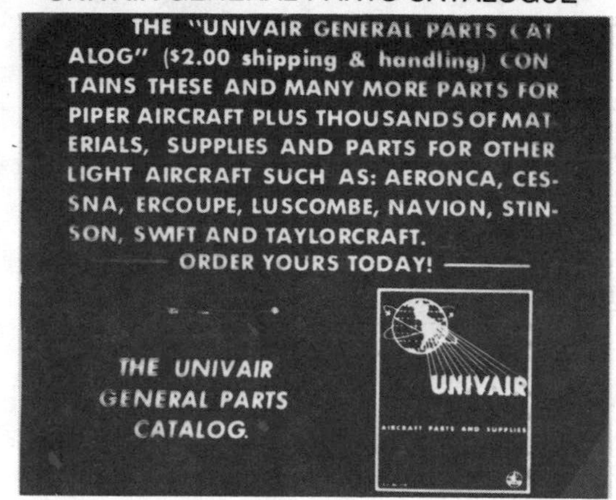

Univair Aircraft Corporation
Route 3
Box 59
Aurora, Colorado 80011
(303) 364-7661

$2 catalog listing aircraft parts and supplies, such as metal cowling, wooden propellers, structural components, wings.

U.S. Industrial Tool & Supply Company
13541 Auburn
Detroit, Michigan 48223
1-800-521-4800

Twenty-page free catalog or 72-page $1 catalog listing aircraft tools, testing equipment and accessories for do-it-yourselfers.

U.S. Department of Transportation
Federal Aviation Administration
800 Independence Avenue SW
Washington, D.C. 20591

Guide to Federal Aviation Administration publications and list of FAA advisory circulars, of which most are free.

Wag-Aero, Inc.
Box 181
Lyons, Wisconsin 53148
(414) 763-9586

Free catalog listing replacement parts and supplies for the homebuilder—antennas, avionics equipment, gas tanks, exhaust systems.

Walnut Products Company
1316 Palmer Avenue
Camarillo, California 93010
(805) 484-5447

Free booklets that explain how to use electronic calculators in the cockpit, plus other pilot supplies.

MOVIES

Aero/Space Visuals Society
2500 Seattle Tower
Seattle, Washington 98101
(206) 624-9090

The Society offers for sale or rental "Threshold: The Blue Angels Experience" in videotape or film. Film is rated G and runs 89 minutes. It is available in videotape in ½" V.H.S., ½" Beta or ¾" U-Matic. Film is either 16 or 35 mm. Charge cards are accepted.

National Cinema Service
P.O. Box 43
Ho-Ho-Kus, New Jersey 07423

The National Cinema Service offers a comprehensive list of all brand new 16 mm. sound features and short subjects for sale. All films are in the public domain. Send for a complete list.

TRAVELING IN YOUR AIRPLANE

From the U.S. Government, from state governments, and from private publishers come publications to make your traveling easier and more pleasurable.

Aeronautical Charts and Related Publications
Distribution Division (C44)
National Ocean Survey, NOAA
Riverdale, Maryland 20840

The U.S. Government publishes a free catalog listing VFR and IFR charts.

Jeppesen Sanderson
8025 East 40 Avenue
Denver, Colorado 80207

Jeppesen Sanderson prints a free catalog listing the IFR charts they publish.

International Flight Operations
AOPA
Box 5800
Washington, D.C. 20014

The International Flight Operations Department, a service of the Aircraft Owners and Pilots Association, provides international flight planning, including computerized jet routes, recommended routes worldwide, up-to-date political information at each stop through daily liaison with U.S. State Department, diplomatic flight clearances, ground aircraft and passenger handling, domestic and international charts, special flight reports. Membership fee: $29.

State governments offer help for a pilot and passenger, including charts, airport directories, newsletters, and maps. Following is an alphabetical listing.

Alabama
State of Alabama
Department of Aeronautics
11 South Union Street
State Highway Building
Montgomery 36104

Free aeronautical chart.

Arizona
Arizona Department of Transportation
Aeronautics Division
205 South 17th Avenue
Phoenix, Arizona 85007

Free aeronautical chart, and airport directory for Arizona residents who pay state and local aircraft fuel taxes.

Arkansas
State of Arkansas
Division of Aeronautics
Adams Field-Old Terminal Building
Little Rock 72202

Free aeronautical chart.

California
Department of Transportation
Division of Aeronautics
1120 N. Street
Sacramento 95814

Free aeronautical chart.

Connecticut
State of Connecticut
Department of Transportation
24 Wolcott Hill Road
P.O. Box Drawer A
Weathersfield 06109

Free airport directory and aviation bulletin.

Delaware
Delaware Transportation Authority
P.O. Box 778
Dover 19901

Safety seminars, airport master planning, airport safety aids, assistance for obstacle removal.

Georgia
State Highway Department of Georgia
Division of Highway Planning
No. 2 Capitol Square
Atlanta, Georgia 30334

Free aeronautical chart.

Hawaii
State of Hawaii
Department of Transportation
Air Transportation Facilities Division
Honolulu International Airport
Honolulu 96819

Free airport directory and flying safety manual.

Illinois
Illinois Department of Transportation
Division of Aeronautics
Capital Airport
North Walnut Street Road
Springfield 62705

Free aeronautical chart, airport directory, and information on aviation seminars for Illinois residents who have registered their airman certificate with the Division and paid the annual $5 fee. For nonresidents and others, they can receive copies of the publications by submitting $5 for the directory and $2 for the chart, payable to the Illinois Division of Aeronautics. They can, however, receive copies of "Illinois Aviation," which lists future seminars.

Indiana
Aeronautics Commission of Indiana
Room 1025
State Office Building
100 North Senate Avenue
Indianapolis 46204

Chart for $2.50.

Iowa
Department of Transportation
Aeronautics Division
State Capitol
Des Moines 50319

Free aeronautical chart, airport directory, and corrections on airport directory.

Kansas
Division of Aviation
Kansas Department of Transportation
State Office Building
Topeka 66612

Free aeronautical chart and airport directory.

Kentucky
Commonwealth of Kentucky
Department of Transportation
Frankfort 40601

Nothing available currently.

Maryland
State Aviation Administration
P.O. Box 8755
Baltimore-Washington International Airport 21240

Aeronautical chart, airport directory, newsletter, all free for state (or former state) pilots only.

Massachusetts
Massachusetts Aeronautics Commission
Boston-Logan Airport
East Boston 02128

Free airport directory.

Michigan
Department of State Highways and Transportation
Aeronautics Commission
Capital City Airport
Lansing 48906

Free airport directory for registered Michigan aircraft owners upon request on previous year's registration form and $3 for others; $1 fee for aeronautical chart. Also available is the newsletter *Michigan Aviation*.

Minnesota
Documents Division
117 University Avenue
St. Paul 55155

$1 charge for aeronautical chart, $2 charge for airport directory.

Mississippi
Mississippi Aeronautics Commission
P.O. Box 5
Jackson 39205

Free aeronautical chart.

Missouri
State of Missouri
Department of Transportation
P.O. Box 1250
Jefferson City 65101

Free aeronautical chart and airport directory.

Montana
Aeronautics Division
P.O. Box 5178
Helena 59601

$2 each for aeronautical chart, airport directory, and newsletter annual subscription.

Nebraska
Department of Aeronautics
Nebraska Aeronautics Commission
Box 82088
Lincoln 68501

Free airport directory for registered Nebraska pilots. For others, $2 charge.

New Hampshire
Aeronautics Commission
Municipal Airport
Concord 03301

Free airport directory and quarterly newsletter.

New Jersey
Department of Transportation
Division of Aeronautics
1035 Parkway Avenue
Trenton 08625

Free airport directory, newsletter called *Flight Log*, and aviation fact sheet.

New Mexico
Department of Aviation
P.O. Box 579
Santa Fe 87501

Free aeronautical chart, brochure describing New Mexico, a road map and a calendar of events.

New York
Department of Transportation
1220 Washington Avenue
State Campus
Albany 12232

Free *Digest of New York State Laws Affecting Aviation*.

North Carolina
Department of Transportation
Division of Aeronautics
Highway Building
P.O. Box 25201
Raleigh 27611

Aeronautical chart, advisory charts of military activity with the state, for free, and a *Skyways* newsletter.

North Dakota
State of North Dakota
Aeronautics Commission
Box U
Bismarck 58505

Free airport directory and aeronautical chart.

Ohio
State of Ohio, Division of Aviation
Ohio State University Airport
2829 W. Granville Road
Worthington 43085

Free Division of Aviation catalog.

Oregon
Oregon State Aeronautics Division
Department of Transportation
3040 25th Street
Salem 97310

Free aeronautical chart.

Pennsylvania
Commonwealth of Pennsylvania
Department of Transportation
Bureau of Aviation
Capital City Airport
New Cumberland 17070

Free aeronautical chart.

Rhode Island
State of Rhode Island
Division of Airports
Theodore Francis Green State Airport
Warwick 02886

Free information on state airports.

South Carolina
South Carolina Aeronautics Commission
P.O. Box 1769
Columbia 29202

Free aeronautical chart.

Texas
Texas Aeronautics Commission
P.O. Box 12607
Capital Station
Austin 78711

Free airport advisory, farm and ranch airstrips pamphlet, sport aircraft builders guide, airport directory, Texas airport system, film library catalog, Mexico flight manual, and newsletter.

Utah
Department of Transportation
Division of Aeronautics
Salt Lake City International Airport
135 North 2400 West
Salt Lake City 84116

Free chart and airport directory.

Virginia
Commonwealth of Virginia
State Corporation Commission
Division of Aeronautics
4508 South Laburnum Avenue
P.O. Box 7716
Richmond 23231

Free chart, airport directory, *Fly to Virginia* brochure, aviation newspaper, and newsletter.

Washington
Department of Transportation
Division of Aeronautics
8600 Perimeter Road
King County International Airport
Seattle 98108

Washington *Pilot's Guide* is sent to pilots registered with the State of Washington.

Wisconsin
Department of Transportation
Document Sales
3617 Pierstorff Street
P.O. Box 7713
Madison 53707

Free aeronautical chart. County maps available for a small fee, as are plastic wall map of Wisconsin and additional maps and documents.

Wyoming
State of Wyoming
Aeronautics Commission
200 East 8th Avenue
Cheyenne 82002

Free aeronautical chart and airport directory.

Some additional guides for traveling are listed below, including several books:

Airport Restaurants
Box 81
West Islip, Long Island
New York 11795

Airport Restaurants & Tour Guide by Ray Turner, biennial magazine lists descriptions of nearly 200 airport restaurants in 46 states. Sold at airports, flight offices, and by mail. $2 per copy. Rating system goes as high as four props.

Pathfinders
Box 11950
Reno, Nevada 89510

Airports of Mexico and Central America by Arnold Senterfitt, 560 pages with mid-year updates, lists measured runways, places for fueling, courses and airways. $24.95.

Airpower Museum
Route 2, Box 172
Ottumwa, Iowa 52501

Airpower Museum located on Antique Airfield, Blakesburg, Iowa has 40 planes on exhibit. The museum store sells coffee cups, necklaces, decals, hats, shirts, cards with logo, tape recordings.

Atlantic Aviation Corporation
Greater Wilmington Airport
P.O. Box 1709
Wilmington, Delaware 19899

Many companies offer ground services, including hangarage, ground facilities (pilot lounges), services-catering, cleaning, courtesy car and rentals, hotel and office reservations, and charters.

Buckingham Air Park
Rt. 1—Box 507
Ft. Myers, Florida 33905
(813) 694-0152

Buckingham Air Park is a fly-in air park with several houses already built and several under construction. The runway stretches 3,000 feet. Contact Charles Braden for more information.

Experimental Aircraft Association Museum
11311 W. Forest Home Avenue
Franklin, Wisconsin 53132

Houses nearly 190 aircraft. Open daily from 8:30 a.m. to 5 p.m. and Sundays from 11 a.m. to 5 p.m.

Dogwood Press
P.O. Box 265
Pinedale, California 93650

Fly-N-Eat, A Guide to Airport Restaurants in California by Steve M. Aldridge.

6 / For the Flying Enthusiast

COLLECTIBLES AND GIFTS

You don't have to be a pilot to sport a belt buckle with an airplane cast in brass or silver, or don a flight suit or polar vest. You may, in fact, have no desire to travel 36,000 feet above the ground. But many companies have discovered that the aviator look is catching on, and may become the sequel to the *Urban Cowboy* fashion trend. Some of the companies manufacture or distribute a single item, such as a watch able to track any event to 1/100th second accuracy. Other companies put forth catalogs brimming with clothing to outfit you from head to toe—goggles, helmets, T-shirts, gloves, wings and patches, ties, shoes, jewelry, and electronic equipment such as scanners.

Following are some high-flying ideas. First are collectibles and gifts, and then clothing and jewelry.

ELECTRA COMPANY

The Bearcat scanners by Electra range from the Bearcat 12, which monitors 10 channels over five bands, to the Bearcat 250, which gives pilots 50 channels. Bearcat also offers antennas.

Bearcat Scanner

An example of the scanners is the Bearcat 220, which covers seven bands, low and high VHF, UHF, UHF-government and UHF-T, two meter amateur (Ham), and aircrafts. Up to 20 frequencies can be scanned, or frequencies can be arranged into two banks of ten frequencies each, allowing the listener to choose the bank of most interest. The 220 features normal search operation where frequency limits are set and the scanner searches between them, and it also searches all marine or aircraft frequencies by pressing a single button. The frequencies are stored in memory so no programming of them is required. Other features include direct channel access for going directly to a desired channel; patented selective scan delay so scanning on desired channels is delayed one second after end of a transmission; and scan speed control, which allows scanning and searching at four or eleven channels per second.

THE COMPLETE BEARCAT® LINE OF SCANNERS

Bearcat® Model	One-Four	6	Hand Held	Four-Six	III	8	12	210	220	250
Features										
Interchangeable modules	—	—	—	—	yes	—	—	—	—	—
Track tuning	yes	yes	—	yes	yes	yes	yes	yes	yes	yes
Single antenna	yes	yes	yes	yes	yes	yes	yes	yes	yes	yes
Single man./scan switch	—	—	yes	yes	yes	yes	yes	—	—	—
Light emitting diodes	yes	yes	yes	yes	yes	yes	yes	—	—	—
Decimal display	—	—	—	—	—	—	—	yes	yes	yes
Frequency Bands										
30-50 mHz low VHF	yes	yes	yes	yes	yes	yes	yes	yes	yes	yes
118-136 mHz aircraft	—	—	—	—	—	—	—	—	yes	—
146-148 mHz "ham"	—	yes	—	—	yes	yes	yes	yes	yes	yes
148-174 mHz high VHF	yes	yes	yes	yes	yes	yes	yes	yes	yes	yes
420-450 mHz 75 CM "ham"	—	—	—	—	—	—	—	yes	yes	yes
450-470 mHz UHF	yes	—	yes	yes	yes	yes	yes	yes	yes	yes
470-512 mHz UHF-T	—	—	yes	yes	yes	yes	yes	yes	yes	yes
Specifications										
Bands	1	2	2	4	1 or 2	5	5	6	7	6
Channels	4	6	4	6	8	8	10	10	20	50
Power	AC	AC	DC	DC	AC/DC	AC	AC/DC	AC/DC	AC/DC	AC/DC
Audio output (rms)	0.5W	1.5W	0.3W	0.3W	2.5W	2.0W	2.0W	2.0W	2.0W	2.0W
Operation	crstl.	crstl.	crstl.	crstl.	crstl.	crstl.	crstl.	syn.	syn.	syn.

Electra Company
Division of Masco Corporation
300 East County Line Road
Cumberland, Indiana 46229

Bearcat Scanner

Flame Engineering Portable Heater

FLAME ENGINEERING, INC.

Flame Engineering manufactures a portable heater, available for either 12 volts DC or 110 volts AC operation. The liquid vaporizer torch is designed to be used with either liquid or vapor LPG. It can melt snow, heat asphalt, heat airplane engines, thaw frozen pipes. Operating instructions are simple, labeled on the blower housing.

Also available are vapor torches, liquid torches, liquid spray torches, dollies, torch kits, hoses, pressure gauges, pipes, LPG cylinders and hitches.

Flame Engineering, Inc.
Box 577
LaCrosse, Kansas 67548

HARVEY PARK RADIO

Harvey Park Radio sells scanners, antennas, and other airplane accessories from some of the best sources—Fanon Courier, Regency, Bearcat, RCA. Write for the catalog.

Harvey Park Radio
Box 19224
Denver, Colorado 80219

HOWARD PRODUCTS CO.

The Howard ratchet set can be used wherever limited space prevents the use of conventional wrenches. Known as the Thumb Wheel, the item

Howard Products Rachet Set

can fit into the palm of the hand or between the thumb and forefinger. Each set contains one reversible ratchet, standard sockets, one extension, and a self-sealing carrying pouch.

Each set includes: thumb wheel ratchet, 2″ extension, 3/16″ socket, 1/4″ socket, 5/16″ socket, 3/8″ socket, 7/16″ socket, 1/2″ socket.

Howard Products Co.
P.O. Box 57246
Dallas, Texas 75207

KAYBRO SALES CO., INC.

Kaybro Sales is the exclusive distributor to the aviation community for "Snoopy's Dream Machine," a motorized mobile.

Snoopy and Woodstock, mounted on a bright yellow flying dog house, battle it out with the evil Red Baron in his red triplane. With guns blazing and props spinning, they circle, each searching for the advantage. No "toy," but the mobile is great fun for big and little Snoopy fans. It makes the perfect

"Snooopy's Dream Machine"

gift for anyone and is a decorative accessory for home or office.

Kaybro also offers a series of authentic scale replica airplanes die-cast in metal, finished with official colors and decals. All have retractable landing gear, rolling wheels, and spinning propellers. The models are six to the assortment. The assortments offered are: commercial airline assortment, Douglas DC-10, DC-9, Lockheed L-1011 Tristar, Boeing's 707, 727 and 747; modern fighter assortment, F-11F, F-104F, F-4E Phantom, Saab Draken, MIG-21, and BAC Jaguar; World War II assortment, A-20J Havoc, Spitfire V, ME 262A Sturmvogel, A6M-5 Zero, P-47 Thunderbolt, and P-51 Mustang.

With the advent of the launching of the space shuttle, Kaybro has introduced die-cast replicas, in scale, of the space shuttle system. All are carefully detailed and bear the official NASA insignia. They include: shuttle package, removable space shuttle rides piggyback on 747 jumbo jet with retractable landing gear, scale 1:500; space shuttle "Columbia," details include opening cargo doors with removable three-section space lab payload, retractable landing gear, scale 1:196; space shuttle "Orbiter," complete with launching pad and two booster rockets that detach from the fuel tank to simulate lift-off, scale 1:500.

Kaybro Sales Co., Inc.
575 S. Barrington Avenue
Los Angeles, California 90049

Scale Replica Airplanes—Commercial Airline Assortment

Scale Replica Airplanes—Modern Fighter Assortment

Scale Replica Airplanes—World War Two Assortment

Kaybro Model Shuttle Package

1514 Space Shuttle "Enterprise"

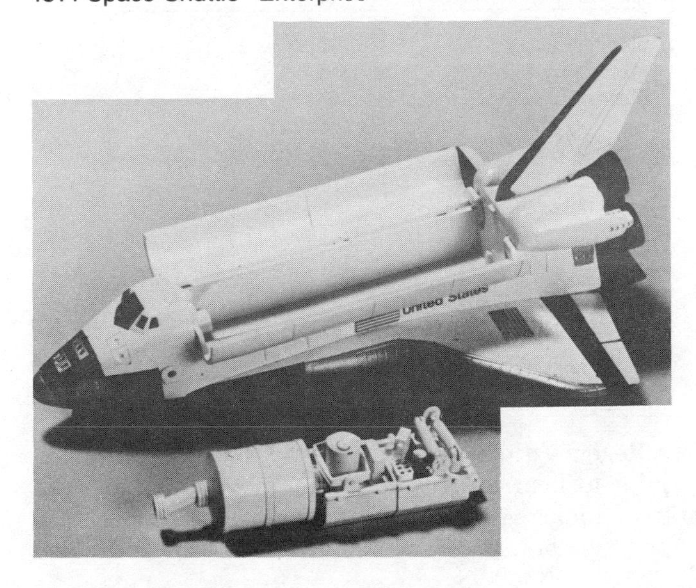

Space Shuttle "Orbiter"

LDB SALES

The Pilot's Flight-Aid by Omnibus has 42 simplified formulas for doing the majority of flight planning calculations quickly and accurately on any four-function pocket calculator. The easy-to-read reference table measures 3½-by-6½ inches. Formulas are listed alphabetically. Its laminated surface resists years of abusive treatment without wrinkling, tearing, staining or fading.

Flying Time clocks include a pen and ink lithograph of the pilot or owner's favorite airplane. The clock dial doubles as the compass rose of the heading indicator. All natural wood case measures 16 inches wide by 6¾ inches high by 2½ inches deep. Plate glass protects clock hands and dial, is solid polished brass. Highest quality West German quartz chronometer keeps precision time—accurate to within three seconds per month. The pen and ink lithograph is of your favorite aircraft. Specify: Cessna 172 Skyhawk, Cessna 310 Twin, Piper Archer, Beechcraft A36 Bonanza, Beechcraft 58 Baron, Piper Navajo. Enquire about availability of other aircraft: military, commercial, classics, aerobatic, gliders.

Multi-Zone clock is for the executive traveler who needs to know in an instant the correct time in several parts of the world. Four precision quartz movements are housed in a wood case that can stand on a table or desk, or hang on a wall. All natural wood case measures 16 inches wide by 6¾ inches high, by 2½ inches deep. Plate glass protects clock hands and dials, is of solid polished brass. Four high-quality West German chronometers keep precision time, accurate to within three seconds per month. Specify any one of the following faces:

Pilot's Flight Aid by Omnibus

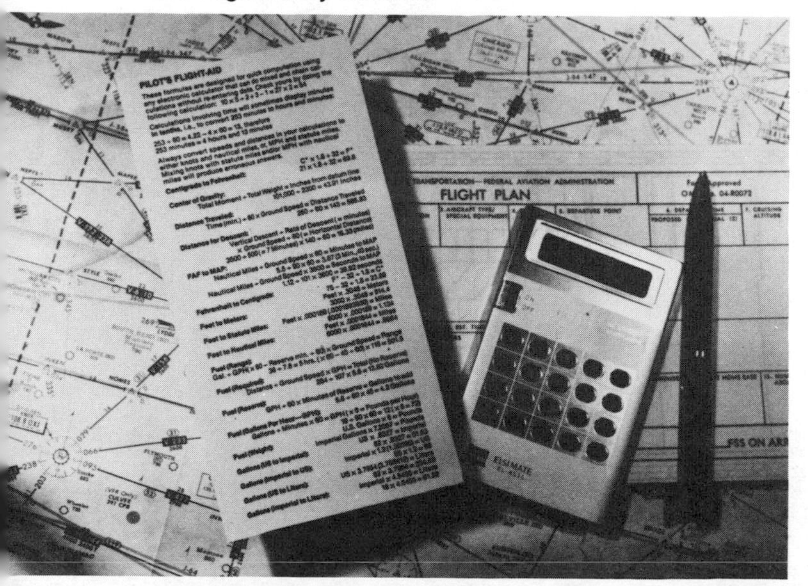

Multi-Zone Clock

Tokyo, San Francisco, Chicago, New York; San Francisco, Chicago, New York, London; Pacific, Mountain Central, Eastern.

LBD
20535 Marathon Court
Olympia Fields, Illinois 60461

REGENCY ELECTRONICS

From Regency Electronics comes flight scan equipment:

Digital Flight Scan puts any civil aircraft navigation or communications frequency within the tip of a pilot's finger. Favorite frequencies in the 16 channels can be stored, or a pilot can search through any part of the aircraft bands for frequencies he or she has never heard in use. There is also a two-channel priority scan function so a pilot can set the emergency frequency, 121.5, plus any other frequency to override all calls coming in on other channels. Other features include battery retaining memory even when unplugged, detachable swivel antenna, individual channel lockouts, and front-mounted speaker. Battery not supplied.

SPECIFICATIONS

Model Number:	ACT-T-720A
Frequency Range:	108–136 mHz
Audio Output:	1 W @ 5% distortion, 2 W max. (8 Ω)
Scan Rate:	approx. 16 chan./sec.
Search Increments:	
108–118 mHz	50 kHz
118-135.975 mHz	25 kHz
Power:	110-130 VAC, 60 Hz, 18 W max.: 11.5-15 VDC, 10 W max.
Listings:	FCC certified, UL-approved

Regency nine-channel Aircraft Scanner allows a pilot to hear aircraft calls on up to nine different frequencies, all automatically. By switching on the priority control, any call coming in on the frequency pilot chooses will automatically override calls on other channels. Supplied with AC and DC power cords, antenna, built-in speaker plus mobile mounting bracket, the Regency weighs four pounds.

SPECIFICATIONS

Model Number	ACT-R-92AP
Frequency range:	
Lo Segment	118-128 mHz
Hi Segment	128–136 mHz
Frequency Separation:	
Lo Segment	10 mHz
Hi Segment	8 mHz
Audio Output:	2 watts (8Ω speaker)
Scan Rate:	approx. 15 chan/sec.
Crystal no.	301-616
Power:	105–130 VAC, 60 Hz; 11–15 VDC; 13 watts, maximum
Listed:	UL Listed, FCC Certified

Regency Electronics, Inc.
7707 Records Street
Indianapolis, Indiana 46226

SKYBUYS

Carry-on luggage eliminates the need for storage in a baggage compartment. Rather, it can be kept in a luggage rack or under a seat on the plane. Made by Skybuys, the Freeloader has separate compartments to prevent wrinkling and crushing of garments and an outside sleeve compartment to hold legal-size papers and documents. In addition, it converts an attaché case or brief case into a roomy travel bag; one size fits all cases. Made of black nylon with saddle trim.

Skybuys
P.O. Box 4111
San Clemente, California 92672

SKYBUYS' Carry-On Luggage

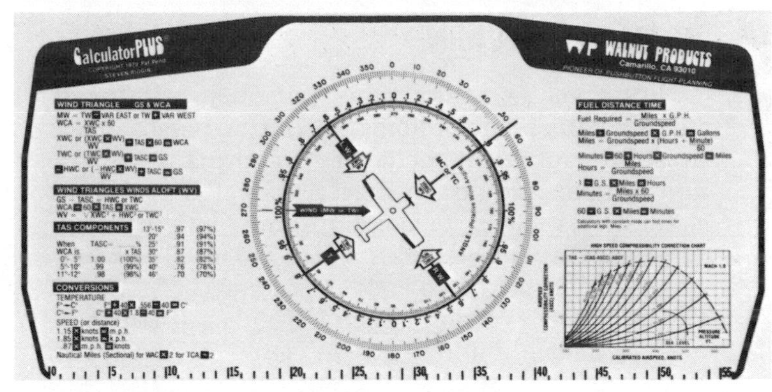

Walnut Product's Calculator Handbook

WALNUT PRODUCTS

Walnut Products manufactures several products for making travel more pleasurable, including a calculator, flight planning safety cassette, calculator handbook, graphic aids package to teach student pilots, private pilot's guide, hook and eye tape to use the calculator in the cockpit, VOR-RNAV and audio visual package consisting of a slide cassette.

Following are more detailed descriptions of some of the Walnut Product items:

Calculator Plus gives pilots a picture of the aircraft and the wind. An optional neckstrap and mounting kit allow easy carrying, instant access. Calculator Plus works also as an ADF computer, traffic pattern computer, holding pattern computer, bearing from landmark computer, plotter and VOR interception headings computer. Neckstrap and cockpit mounting kits are available.

Flight Planning Safety Cassette (see photograph) explains not only the theory but the art of using basic methods and inexpensive calculators.

Graphic Aids package includes wall charts, graphic aids and a manual for students/flight instructors.

Private Pilot's Guide is a flight course based on modern navigation methods. Spanning more than 300 pages, the guide includes illustrations, figures and examples.

Calculator Handbook includes keystroke sequenced formulas.

Audio Visual Package shows students or customers how to use Calculator Plus.

VOR-RNAV program allows a pilot to determine his or her distance immediately plus bearing to any waypoint within the range of two VORs or DME.

Walnut Products
1316 Palmer Avenue
Camarillo, California 93010

WITH YOU IN MIND

With You in Mind offers personalized aircraft items for the plane or home, including personalized aviation napkins in choice of size, airplane, napkin color, ink color and message; personalized aviation note sheets in white bond printed with person's choice of airplane and message in red and black ink; and stained glass airplanes in a choice of styles—Cessna, Piper Warrior, biplane, Beechcraft Duchess, and Aerostar. Photographs of some of the items follow:

1. Personalized aviator's napkins (box of 100)
2. Personalized aviator's note paper (10 pads of 100)
3. Small stained-glass airplanes *
4. Mobil (3 stained-glass airplanes)
5. Aviator's note cards (box of 20)
6. Aviator's greeting cards (assorted) (box of 20)
7. Aviator's Christmas cards (package of 20)
8. Stained-glass airplane (scale mod. Fokker DVII)

* Biplane, low-wing and high-wing

With You in Mind
P.O. Box 405
Dedham, Massachusetts 02026

Stained Glass Airplane Mobile

CLOTHING AND JEWELRY
A-B EMBLEM

With the *Design-an-Emblem Guide,* you can create your own emblem design. All a person needs to do is determine the size and shape of the emblem required, the basic design, colors preferred, and quantity (minimum quantities are shown beside each emblem). Send for free *Design-an-Emblem Guide* and let your imagination go wild.

A-B Emblem
P.O. Box 695
Weaverville, North Carolina 28787

Personalized Aviator's Note Paper

Small Stained Glass Airplane

ACE WILSON COMPANY

Ace Wilson manufactures T-shirts with original aviation-related slogans:

- "Do Not Hand Prop" T-shirt for women only, each $7.95 plus postage.
- "Has Anyone Seen My Fokker" T-shirt for women only, each $5.95 plus postage.
- "If You Ain't a Pilot" T-shirt, each $7.95 plus postage. Quantities of one dozen or more are $4.50 each.
- "Don't Let the Turkeys Keep You Down!" T-shirt, each $7.95 plus postage.
- "The Joy of Sex Doesn't Last Like the Fun of Flying" T-shirt, $6.95 each plus postage.

Two T-shirts, mailed UPS, cost $1.50.

Ace Wilson Company
P.O. Box 2307
Stanford, California 94305

THE AIRPLANE SHOP, INC.

The shop is located on Caldwell/Essex County Airport, 18 miles west of New York City, and serves the hobbyist and collector of World War I and II memorabilia with a varied stock of aviators' helmets, goggles, leather jackets, sheepskin coats and pants, boots, dress and work uniforms, insignia, patches, and wings.

The shop also has a varied selection of aviation-related books and magazines, wood and metal aircraft models along with antique wood and metal propellers, Flying Tiger Blood Chits and hundreds of other aviation collectibles.

The Airplane Shop, Inc.
125 Passaic Avenue
Fairfield, New Jersey 07006

The Airplane Shop

A.T. PATCH CO.

Swiss embroidered emblems by A.T. Patch are loom machine-made or individually machine-made. For a loom emblem, a customer must send to A.T. Patch a sketch or art work of design desired, the size, quantity, and colors. Two-inch emblems must be ordered in multiples of 320; three-inch emblems in multiples of 210; four-inch emblems in multiples of 160. For a hand-machine emblem a sketch or design must be sent, along with colors desired. Customer should measure the longest side of the design to price the emblem. A $5.00 art charge is additional.

Also available from the company are custom decals and removable bumper stickers.

A.T. Patch Co.
Dept 52
R.F.D. 1
Littleton, New Hampshire 03561

AVIREX LIMITED

Avirex Ltd. reproduces leather and sheepskin flying jackets, flight suits, helmets and other apparel and equipment worn and used by pilots and air crews before and during the Second World War. Whenever possible, Avirex obtains the original military patterns, templates, and specifications. The same materials often come from the original World War II suppliers. All items are illustrated in the Cockpit catalog, a division of Avirex, a mail-order service for pilots, military and aviation buffs, motorcyclists, and other outdoor enthusiasts.

A few examples are illustrated here:

A.T. PATCH CO. Embroidered Emblems

Avirex Ltd. Aviator Accessories

The watch offered includes the famous military "Hack" feature of stop second hand for synchronization. Also features stainless steel case, 17-jewel movement, luminous 24-hour dial.

There are many goggles to choose from.

A Blood chit, unframed for sewing on jackets, is available.

A water-resistant jacket, with leather-tanned exterior, natural sheepskin fur, and double-strapped fleece collar. The jacket has sides and single patch pocket, depending upon size.

A nylon flight cloth jacket includes leather tabs and snap straps for oxygen hose retention, left sleeve insignia as originally imprinted on the sleeves of World War II issue flight jackets.

The goatskin-grained leather flight jacket features sheepskin mouton fur collar, contoured pockets, lining, knit waistband and cuffs, and interior map pocket and navigator's instrument pocket under the left pocket flap.

The flight suit for women is made of nylon parachute material with zipper pockets at the chest and waist and also in the legs. There is also a two-way front zipper, a cigarette/pen/pencil pocket on the sleeve and tabs on sleeves, waist, and ankles.

Avirex Flying Jackets, Flying Suits, and Equipment

The flight jacket has real mouton fur collar. Samurai headband is a copy of those worn by Japanese combat and kamikaze pilots in World War II.

Avirex Ltd.
468 Park Avenue South
New York, New York 10016

THE BELT BUCKLE BUSINESS

All of the buckles are manufactured by the "lost wax" method of investment casting in solid brass. Special orders can be cast in sterling silver or 10 karat gold. Subjects include others besides aviation: flags, trucks, animals, recreation, and, in addition, the company also produces T-shirts to match the design of aviation belt buckles.

The Belt Buckle Business
P.O. Box 1395
Norman, Oklahoma 73070

A Selection From THE BELT BUCKLE BUSINESS

THE COCKPIT

The Cockpit is a mail-order service for pilots, aviation buffs, motorcyclists, and all sports enthusiasts. Apparel includes sheepskin flight jackets, hooded parkas, vests, trousers, helmets, sweaters, patches, jewelry, buckles and belts, insignia, aviator glasses, hats, ties, boots, gloves, watches, plus binoculars, clocks, headsets and microphones, knives, and duffle bags.

See Avirex for examples of items offered in the Cockpit catalog.

The Cockpit
Division of Avirex Ltd.
468 Park Avenue South
New York, N.Y. 10016

FLIGHT APPAREL IDENTITY AND CAREER CLOTHING

Flight Apparel offers custom-tailored apparel for aviation, including blazers, jumpsuits, windbreakers, caps, vests, polar jackets, shoes, shirts, shoulder stripes, and slacks, all in a variety of colors, materials, and sizes. Write to them for specifics.

A few examples include:

Permapress flight suit, available for both men and women, features a belt with velcro closure, elastic waistband, and full two-way zipper. The men's flight suit has ten pockets, ladies' suits have fewer pockets. Available in poly/cotton poplin, 100 percent polyester knit or brushed denim and in blue or red colors with Grumman American striping or blue brushed denim and khaki. Price: $21.95

The quilted jacket with hood is Dacron-quilted with nylon shell and lining, full length zipper,

Flight Suits, Quilter Liner Jackets, and Windbreakers from FLIGHT APPAREL IDENTITY AND CAREER CLOTHING

knitted cuffs and two waist pockets. The navy blue exterior reverses to orange for higher visibility. The jacket is machine washable and available in men's, women's and children's sizes. Price: $14.98.

The windbreaker jacket is water-repellant, with a full-length zipper, knit cuffs, and pockets inside and out. The windbreaker comes in either poly/cotton poplin or 100 percent Dacron polyester knit in Grumman American colors of red, blue, and white. Price: $12.98.

Flight Apparel Identity and Career Clothing
Hammonton Airport
P.O. Box 166
Hammonton, New Jersey 08037

FLIGHT SUITS LTD.

Each flight suit has two large zippered chest pockets and two zippered chart pockets on the legs. Long-sleeve suits have a zippered bellows pocket with an attached pen-and-pencil pocket on the sleeve, while short-sleeve suits have a pen-and-pencil pocket only. Two-piece flight suits consisting of a shirt and trousers are also available.

Women's flight suits have the same construction features as men's suits, but have either a belt fastening in the center with Velcro or a tie belt.

All the flight suits and jackets have the same basic cut and fit, but can be individually designed through the use of design options, striping, colors, and fabrics.

The two models include the professional with concealed zippers, padded shoulders, epaulets or chest striping, and the sportster with exposed zippers, sleeve and leg striping, embroidered nametags and epaulets.

The jackets are also available in the two models.

Flight Suits Ltd.
1158 North Marshall
El Cajon, Calif. 92020

A Selection from FLIGHT SUITS LTD.

Fudpucker World Airlines Captains Club Membership Certificates

CAPTAINS CLUB

In recognition of a continuing and fervent interest in the development and future of coal burning air transportation, Fudpucker World Airlines has recognized the following person for membership - and hereby, forthwith, and to wit, do name the following as:

Esteamed Honorable Cap...

entitled to all rights and privileges, ...

thereto, from this day forward.

CAPT. ______________

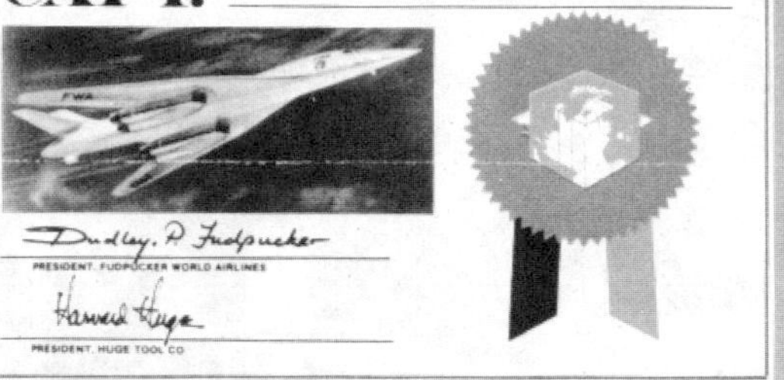

Dudley P. Fudpucker
PRESIDENT, FUDPUCKER WORLD AIRLINES

Howard Huge
PRESIDENT, HUGE TOOL CO.

The coveted Dudley P. Fudpucker "Fudpucker World Airlines Captain's Club Membership Certificates." Just **$5.95.** CAPTAIN'S CERTIFICATE or the STEWARDESS CERTIFICATE. Exclusively Fudpucker, ready for framing. Full size 8-1/2" x 14". Please be sure to specify proper certificate and print name to be inserted, on order blank.

CAPTAINS CLUB
Stewardess Division

In recognition of a continuing and fervent interest in the development and future of coal burning air transportation, Fudpucker World Airlines has recognized the following person for membership - and hereby, forthwith, and to wit, do name the following as:

...norary Crew Member #1,

...ll rights and privileges pertaining

...n this day forward.

Stew. ______________

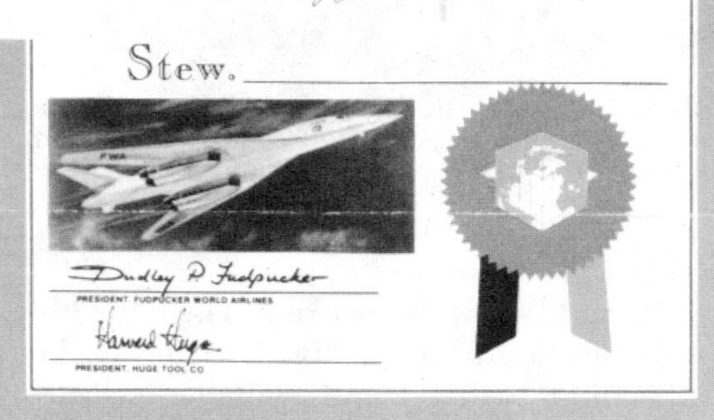

Dudley P. Fudpucker
PRESIDENT, FUDPUCKER WORLD AIRLINES

Howard Huge
PRESIDENT, HUGE TOOL CO.

FUDPUCKER EMBLEMS

Bright colored 4" Swiss embroidered sew-on patches proclaim you're a Fudpucker. Neat trim for a favorite jacket and cap. Selection includes:

A. Fudpucker World Airlines
B. Fudpucker Trucker
C. Fudpucker Van'R
D. Fudpucker Gun Club
E. Fudpucker Jogger

Priced at .... **$5.95 each**

Order on Fudpucker World Airlines order blank.

Selections from "The Fudpucker Catalogue"

FUDPUCKER

The Fudpucker catalog includes T-shirts, nylon jackets, "Captain" caps, emblems and shorts plus assorted paraphernalia with aviation motifs for the home, such as cocktail napkins, beer mugs, luggage tags and playing cards.

Some of the items include:

RT The preferred look of the classic T-shirt trimmed in dark blue. 50% cotton, 50% polyester with reinforced neck. Specify S, M, L, XL

J Solid blue, white trimmed sports shorts. 100% sanforized cotton twill. Covered elastic waistband. Side vents. Specify waist sizes XS, S, M, L, XL

RJ All purpose flannel-lined nylon jacket. Snap button front with raglan sleeves, deep slash pockets, draw-string waist and elastic push-up sleeves. Soil resistant. Navy, specify S, M, L, XL

RC Solid front, mesh back give this cap a racing flair. Sizing strap to fit all.

RL Fashion fitted ladies top for your racy lady. 50% cotton, 50% polyester. Powder blue. Sizes S (32), M (34), L (36)

L Sunburst lemon yellow ladies top with "Fudpucker" in colors emblazoned on the front. 50% cotton, 50% polyester. Sizes S (32), M (34), L (36)

S Thigh length, super soft, slumber shirt. Great for the sauna or beach. V-neck and thigh vents. Specify S, M, L, XL

Fudpucker Emblems are bright-colored, 4″ Swiss embroidered sew-on patches and proclaim you're a Fudpucker. Neat trim for a favorite jacket and cap. Selection includes: A. Fudpucker World Airlines, B. Fudpucker Trucker, C. Fudpucker Van'R, D. Fudpucker Gun Club, F. Fudpucker Jogger.

Fudpucker
Box 67
Hayden Lake, Idaho 83835

INNOVATIVE TIME CORPORATION

The Olympia V is a shock- and water-resistant watch with an ultralight case and scratch-resistant mineral glass lens. Special features include a precision quartz chronograph/stopwatch capable of timing any event with track accuracy to 1/100th second, two separate alarms—the first to sound exactly on the chosen minute and the second to chime every hour on the hour, and a lithium battery with life of approximately three years.

Innovative Time Corporation
350-A Fischer Avenue
Costa Mesa, California 92626

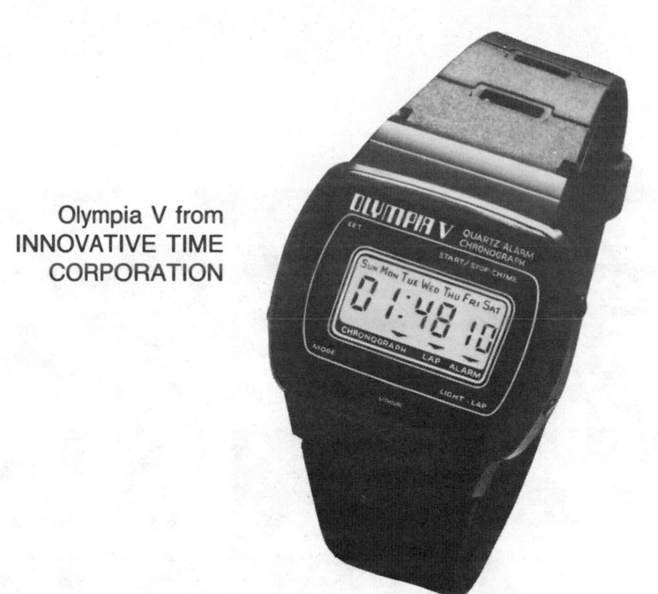

Olympia V from INNOVATIVE TIME CORPORATION

KAYBRO SALES CO., INC.

Kaybro Sales has a line of aviation jewelry, including tie tacks, tie bars, cuff links, charms, necklaces, earrings, and pins for women.

The jewelry is approximately 1-by-1¼ inches in length, except for pins, which are 2 inches long. All styles are available in heavy gold plate or rhodium silverplate. The various series are available in the following styles. Write for complete catalog.

Kaybro Sales Co., Inc.
Post Office Box 24916
Los Angeles, California 90024

SKYBUYS

Skybuys manufactures Skylites, jewelry which can be worn as a necklace, bracelet, or tie tack. Each features a working red warning beacon on a real airplane, which "warns" other persons that an airplane-lover is in sight. Skylites come complete with a chain for the necklace version. All contain a battery that will operate the Skylite for three weeks, but can be disconnected when the Skylite is not being worn.

Skybuys also offers the pilot, passenger, or aviation dreamer shirts, caps, visors, and jewelry with aviation motifs. Wings and epaulets are silk-screened on clothing in wash-proof textile paint.

Chief Pilot golf shirt is 50 percent cotton and 50 percent polyester. Shirt has one pocket, four-button placket, long body for in/out wear. Comes in white only. Sizes: S, M, L, XL.

Chief Pilot T-shirt has a wide band, crew neck with pocket. The shirt is 100 percent cotton. Colors

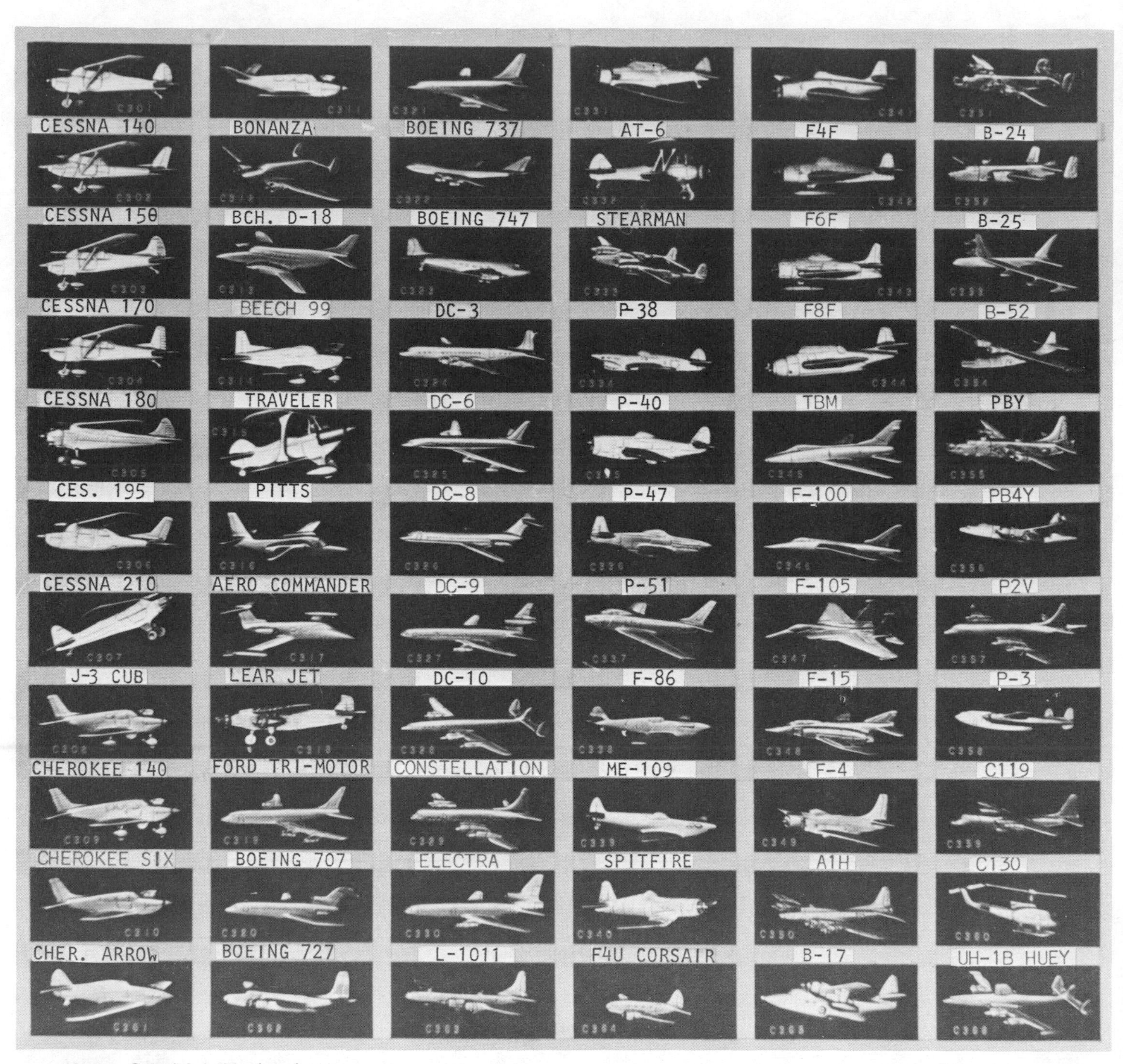

Kaybro Sales' Aviation Jewelry

Kaybro Sales' Aviation Jewelry

Skybuys' Chief Pilot Golf Shirt

Skybuys' Chief Pilot Cap

are tan or light blue with navy blue wings and epaulets. Sizes: S, M, L and XL.

Chief Pilot caps are mesh-back baseball-style caps with long-bill visors and adjustable headbands to fit all sizes. Navy blue wings on white front panels; choice of red, yellow, light blue or navy blue caps.

Chief Pilot sun visors are tennis-style sun visors that carry Chief Pilot wings on front in navy blue. Choice of white, yellow, light blue or red visors.

Four weeks for delivery, Visa and MasterCard accepted.

Skybuys
P.O. Box 4111
San Clemente, California 92672

Split-S Aviation Leather Jacket

SPLIT-S AVIATION

Split-S began by offering leather helmets and goggles and now offers, in addition, silk flying scarves, flying gloves, flight jackets, and more. Illustrated is the deluxe bomber jacket for men in navy.

Split-S Aviation
1050-J Pioneer Way
El Cajon, California 92020

INDEX